AUTONOMY AND NORMATIVITY

'A radical and systematic critique of some of the mainstream thinkers by means of a single unifying principle. This book will contribute very constructively to current philosophical debate.'

Wilfried Ver Eecke, Professor of Philosophy, Georgetown University, USA

'Winfield has a very clear position, namely that normativity must be foundation-free and so self-determining, and he explores the implications of this position with great energy and clarity. It is very helpful to have these essays collected together so coherently in this single volume which will appeal to students and scholars coming from a broadly Hegelian direction, as well as those working on political questions from the perspectives of Kant, Marx, Rawls or Habermas.'

Stephen Houlgate, Professor of Philosophy, University of Warwick, UK

Autonomy and Normativity explores central topics in current philosophical debate, challenging the prevailing post-modern dogma that theory, practice and art are captive to contingent historical foundations by showing how foundational dilemmas are overcome once validity is recognized to reside in self-determination.

Through constructive arguments covering the principal topics and controversies in epistemology, ethics, and aesthetics, *Autonomy and Normativity* demonstrates how truth, right and beauty can retain universal validity without succumbing to the mistaken Enlightenment strategy of seeking foundations for rational autonomy. Presenting a compact, yet comprehensive statement of a powerful and provocative alternative to the reigning orthodoxies of current philosophical debate, Richard Winfield employs Hegelian techniques and focus to object to opponents, and presents a radical and systematic critique of the work of mainstream thinkers including Kant, Rawls, Husserl, Habermas and others. The ramifications for the legitimation of modernity are thoroughly explored, in conjunction with an analysis of the fate of theory, practice and art in the modern world. This book offers an invaluable resource for students of both analytic and continental philosophical traditions, and related areas of law, social theory and aesthetics.

In memory of my brother-in-law,
Siddhartha Gautam

Autonomy and Normativity

Investigations of Truth, Right and Beauty

RICHARD DIEN WINFIELD

Routledge
Taylor & Francis Group
LONDON AND NEW YORK

First published 2001 by Ashgate Publishing

Reissued 2018 by Routledge
2 Park Square, Milton Park, Abingdon, Oxon, OX14 4RN
711 Third Avenue, New York, NY 10017, USA

Routledge is an imprint of the Taylor & Francis Group, an informa business

A Library of Congress record exists under LC control number: 00046896

ISBN 13: 978-1-138-70679-8 (hbk)
ISBN 13: 978-1-315-20162-7 (ebk)

Contents

A Note on the Text

Those chapters that were previously published are listed below with the corresponding publication credits. The respective journals and presses are thanked for granting permission to include these studies, which are here integrated with certain alterations.

Chapter One: "Freedom from Foundations: The Normativity of Autonomy in Theory and Practice", **Jadavpur Journal of Philosophy**, Vol. 4, No. 1, Fall 1992, pp. 1–27.

Chapter Two: "Hegel's Remedy for the Impasse of Contemporary Philosophy", **Reason Papers**, Vol. 16, Summer 1991, pp. 115–132.

Chapter Three: Review of William Maker's *Philosophy without Foundations: Rethinking Hegel*, **The Owl of Minerva**, Vol. 28, No. 1, Fall 1996, pp. 100–108.

Chapter Four: "Concept, Individuality and Truth", **Bulletin of the Hegel Society of Great Britain**, Nos. 39/40, 1999, pp. 35–46.

Chapter Five: "Space, Time and Matter: Conceiving Nature without Foundations", in **Hegel's Philosophy of Nature**, edited by Stephen Houlgate (Albany: State University of New York Press, 1998), pp. 51–68.

Chapter Six: "Ethical Community without Communitarianism", **Philosophy Today**, No. 2, 1996, pp. 310–320.

Chapter Eight: "With What Must Ethics Begin? Reflections on Benson's Account of Property and Contract", **Cardozo Law Review**, Vol. 11, No. 3, February 1990, pp. 537–548.

Chapter Nine: "Unity in the Common Law?", **The Canadian Journal of Law & Jurisprudence**, Vol. IX, No. 2, July 1996, pp. 411–422.

Chapter Ten: "Friendship, Family and Ethical Community", **The Philosophical Forum**, Vol. 28, No. 4 and Vol. 29, No. 1, Fall 1997–Winter 1998, pp. 1–20.

Chapter Twelve: "Should the Economy be Democratized?", **Southeastern Political Review**, Vol. 23, No. 4, December 1995, pp. 689–702.

Chapter Thirteen: "Relativism and Democracy", **Jadavpur Journal of Philosophy**, Vol. 10, No. 1, Fall 1998, pp. 1–12.

Chapter Fourteen: "Post-Colonialism and Right", in **Hegel's Philosophy of Right**, edited by Robert Williams (Albany: State University of New York Press, 2001).

Chapter Fifteen: "The Individuality of Art and the Collapse of Metaphysical Aesthetics", in **American Philosophical Quarterly**, Vol. 31, No. 1, January 1994, pp. 39–51.

Chapter Sixteen: "Hegel, Romanticism, and Modernity", **The Owl of Minerva**, Vol. 27, No. 1, Fall 1995, pp. 3–18.

Chapter Seventeen: "The Challenge of Architecture to Hegel's Aesthetics", in **Hegel's Aesthetics**, edited by William Maker (Albany: State University of New York Press, 2000), pp. 97–111.

Introduction

Autonomy and normativity have had an often beleaguered association in the annals of theory, practice, and art, but seldom has their link been suppressed. From the beginning, philosophy has understood its pursuit of truth to depend upon the autonomy of reason. Without reason's liberty to question everything, philosophical inquiry could never escape the hold of prevailing authority and a relativizing subservience to whatever claims that authority privileges. Moreover, without the positive freedom to generate ideas independently of antecedent dogma, philosophy could never advance from conditioned opinion to truths unconditioned by bondage to unexamined givens. Whenever the legitimacy of conduct has become an issue, freedom has equally had to be enlisted, first, as the constituent gateway to the sphere of responsible action circumscribing right and wrong, and second, as the conduit from enslavement under ruling convention to the enactment of practices that are as they ought to be. Finally, for beauty to be evaluated in its own right, shorn of magical, sacred, or prosaic concerns, the necessities of given form and antecedent rules have had to make way for autonomy to accommodate the individuality, originality and creativity pervading the aesthetic fit of meaning and configuration.

At each turn, the demands of normative justification require a liberation from the given as well as a positive autonomy to deliver a content independent of external foundations. Neither truth, nor right, nor beauty can be defined by anything lacking the normativity they each constitutively possess. Yet if truth cannot be determined by what lies outside the truth, if what is right cannot be prescribed by what is not itself right, and if beauty cannot be extracted from what lacks aesthetic worth, that is, if, generally, what is normative cannot be defined by what is normatively neutral, normativity seems to revert to an empty tautology where whatever claims validity for itself has nothing to offer in its own support than the self-identity that any arbitrary assertion might share. The alternatives seem starkly bleak: if normativity gets rooted in some distinct foundation, that foundation can never satisfy the standard of normativity that it endows by founding what derives from it. As the antecedent ground of what counts as valid, the privileged foundation would have to be its own ground to enjoy the normativity it establishes. Otherwise, the foundation's own character would be a dogmatic given that cannot meet its own requirement of resting upon the privileged ground from which validity is supposed to derive. Yet if the foundation does rest upon itself, it ceases to be the foundation of something other than itself, eliminating the entire relation of ground and grounded by which its own privileged foundational role is defined. On the other hand, if normativity cannot be based upon something lying outside itself, it seems condemned to an arbitrary content, whose presumed privilege can be challenged by any competitor, insofar as each and every discernible candidate can claim an individuality making it a law unto itself. Normativity would seem reduced to the sheer positivity of residing in whatever is put forward as exclusively privileged.

Although the horns of this dilemma have appeared insurmountable to many, its formulation already contains the key to its own solution. In order for any putative foundation of normativity to satisfy its own standard for validity, it must ground itself. Accordingly, whenever foundational justification attempts to achieve internal consistency, the ground of normative validity becomes impelled to determine itself, thereby eliminating the distinction between prior foundation and what the foundation endows with validity. In effect, normativity becomes equivalent to self-determination, for only what is what it determines itself to be escapes the dependence upon some ground that renders justification problematic.

Although this identification of freedom and normativity results from foundationalism's own internal imperatives, it is hardly conditioned by any initial attachment to foundational justification. There simply is no other alternative for normative justification, for if what is valid is not self-determined, its validity is determined by something else, reintroducing the ground-grounded relationship defining foundationalism. Validity then once again derives from a separate legitimating factor, which can only meet its own validity requirement by grounding itself and thereby reverting to self-determination. Opposition to the normativity of freedom is thus refutable without appeal to any separate commitment, demanding justification for itself. Not only does all opposition collapse on its own terms, but this collapse leads justification directly to autonomy. No positive claim of any sort need be supplied to provide this transition. Of course, if the move from the collapse of foundational justification to the normativity of autonomy depended upon any connecting principle this mediation would subvert the defining independence of freedom.

What saves self-determination from collapsing into the arbitrary assertion of any and every given is that what is self-determined cannot have any predetermined content. After all, if freedom did have any antecedent nature, it would possess a character that it had not determined by itself. Contents that are arbitrarily asserted are not self-determined precisely because they owe their presence to the arbitrary process of selection that has put them forward. Although arbitrariness and freedom are commonly conflated by the thinking that cannot free itself of foundations, the capacity of choice is not autonomous precisely because whenever it posits arbitrary contents, what gets determined are the choices, rather than the agency of choice, which must already be at hand if any stipulation is to be made. Hence, just as arbitrary contents are determined by an agency separate from themselves, that agency determines not itself, but choices presupposing its faculty of choice, which remains the same no matter what issues from it. By contrast, what is self-determined cannot have an arbitrary content insofar as freedom cannot owe its character to what any given agency has imparted to it.

These considerations might appear to be of merely theoretical significance, but they have a decisive bearing upon the most important practical struggles of modern times.

As much as modernity has been a battleground of competing projects of civilization, it has distinguished itself from other periods by giving rise to the venture of erecting a civilization in which what alone counts as valid in theory, conduct and art is what calls into question the authority of all given tradition and conforms to rational autonomy. By contrast, pre-modern tradition accepts the sanctity of some given heritage that it privileges, allowing that particular foundation to enjoy unchallenged legitimacy. Post-modern ideology, for its part, denies that reason, conduct and art can escape the hold of given foundations, while regarding each grounding framework as a contingent historical convention. Some post-modernists would like to observe an ironic neutrality to all such conventions, but the more resolute path would be to recognize their contingent particularity for what it is by embracing a will to power, advancing particular norms with the brazen consistency of a play to impose arbitrary, perspectival values upon others. Either way, when post-modernists deny that any one foundation has universal validity, they still commit the foundationalist blunder of making a global diagnosis of the historically conditioned fate of reason, conduct and art. Weaving a grand narrative of their own that should undercut the universal validity of any theory or practice, post-modernists can never legitimate the validity of their own hegemonic understanding of the contingent conventions on which they ground every assertion of validity. Either the post-modern diagnosis itself issues from contingent foundations, rendering its assertions as ideologically relative as any competing views, or it claims an unconditioned validity contradicting its own historical conditioning of theory and practice. Although post-modernism may be correct in unmasking the arbitrariness of foundational justification, its inability to conceive of any alternative leaves it prey to the very difficulties it exposes. Because post-modernism is blind to the tie between autonomy and normativity it is equally blind to the truly distinctive character of modern times.

The tie between autonomy and normativity is precisely what enables the project of modernity to escape the dilemmas of foundational justification and command an authority that no pre-modern nor post-modern civilization can match. This unique standing of modernity has been tarnished by the prevailing dogma that the legitimation of modernity is wedded to the Enlightenment Project of providing foundations for rational autonomy. Although Enlightenment thought certainly endorses the sovereignty of autonomous reason and freedom, it misconstrues their content by seeking foundations for what cannot possibly have any. Consequently, the project of modernity is hardly impugned when postmodernism exposes the foundational dilemmas undermining the Enlightenment's effort to fashion objectivity from the structure of consciousness or language, to construct justice from a privileged choice procedure, or to derive beauty from the process of aesthetic reception.

The following investigations of truth, right and beauty provide a post-Enlightenment tonic to the reigning incomprehension of the unity of autonomy and normativity and of the foundation-free legitimacy of modernity.

The five chapters of Part I investigate how truth can be conceived without foundations by duly realizing the autonomy of reason. Chapter One presents the basic argument for the normativity of autonomy in theory and practice by first delineating the dilemmas plaguing the fundamental forms of foundationalism and then sketching how their problems are overcome when thought and action become self-determined. Chapters Two and Three explore how foundational problems continue to undermine the most prominent schools of contemporary Analytic and Continental epistemology and how Hegel's *Phenomenology of Spirit* and *Science of Logic* pioneer a feasible alternative strategy for developing philosophy without foundations. Chapter Four investigates how the autonomy and individuality of conceptual determination enable reason to conceive truth without falling prey to the traditional problems that would relegate philosophy to a mortuary of empty universals. In substantiation of this capability, Chapter Five conceives the most rudimentary features of nature by relying upon an autonomous conceptual development that avoids all appeal to foundations.

The constitutive role of freedom in normative conduct is addressed in all its key dimensions in Part II. Chapters Six and Seven clear the way by retrieving ethical community from the deformations of communitarian thought and outlining how ethics becomes a philosophy of right, conceiving freedom in its three basic forms: property right, morality, and ethical community. Chapter Eight explores the way in which property relations comprise the basic form of freedom incorporated by, but not grounding every other sphere of ethics, whereas Chapter Nine examines how the relation of property right to other more complex forms of freedom allows the common law to retain its unity. Chapters Ten and Eleven turn to family values, examining first how the family should be reconstructed as an institution of freedom, distinct from property relations, morality and friendship, and then how family law requires further alteration to uphold the valid relations of family right. Chapter Twelve addresses the character of economic freedom and whether economic democratization is either a necessary or admissible element in a just economy. The legitimacy of political democracy is next examined in Chapter Thirteen, rescuing the normativity of self-government from the ethical indifference of relativism. Finally, Chapter Fourteen examines the normative challenge of post-colonialism in respect to the ethical disparities between modernity, pre-modernity, and post-modernity.

Analogous connections between autonomy and normativity in aesthetics are put in perspective in Part III, which addresses all three levels of aesthetic theory: the conception of what pertains to art in general, the analysis of the particular styles of art, and the theory of the individual arts. Chapter Fifteen paves the way for systematic aesthetics by exposing how the mimetic theory of classical aesthetics succumbs to foundational dilemmas that bar comprehension of the individuality and autonomy constitutive of the aesthetic unity of meaning and configuration central to beauty, artistic creation, and aesthetic judgment. Chapter Sixteen proceeds beyond the concepts of art in general to investigate the particular character of romanticism, exploring to what degree Hegel's theory of romantic style succeeds in comprehending the shape of art in modernity. Chapter Seventeen addresses the challenge architecture poses to aesthetics, engaging once more in critical debate with the analysis of Hegel's *Aesthetics*, which represents the first and foremost attempt to conceive fine art with due attention to how autonomy underlies the aesthetic unity of meaning and configuration. The specific problems of architecture's functionality and non-representational abstraction are examined in respect to how they can be vehicles for realizing the general features of art, as well as the specific unities of shape and significance distinguishing particular artistic styles.

The ramifications of the normativity of autonomy for truth, right and beauty are drawn together in the conclusion, which focuses upon the legitimation of modernity in connection with the fate of theory, practice, and art in modern times.

I
Truth

One

Freedom from Foundations: The Normativity of Autonomy in Theory and Practice

The Problem of Foundations in Theory and Practice

The fate of freedom stands very much at a crossroads. On the one hand, the call for self-determination has gathered renewed momentum, penetrating all spheres of conduct. At the most rudimentary level of right, the entitlement of ownership has been claimed for all rational agents, excluding slavery once and for all. At the level of moral reflection, the right to an independent conscience has gained ever widening ground, freeing personal conduct from the yoke of elders, clergy, and all other privileged authority. In the household, the family has become increasingly liberated from external restraints that prevent adults from marrying who they wish and from freely co-managing household affairs, irrespective of race, caste, nationality, religion, and, most recently, gender and sexual preference. At the social plane, the clamor grows for universally instituting a civil society, wherein individuals can pursue interests of their own choosing in reciprocity with others, enjoying an equal economic opportunity under conditions of a publicly regulated market. And finally, at the summit of politics, constitutional democracy continues to widen its authority, even as it grapples with the problems of preventing political freedom from being subverted by privileged social interests.

Yet just as the institutions of freedom become more and more the universal goal of practice, the very legitimacy of freedom has come under a growing and insidious challenge. This challenge resides not so much in the backwater efforts of ethnic nationalists and religious fundamentalists to unify the remnants of collapsing empires under principles extraneous to self-determination. The more insidious challenge finds its greatest voice in the heart of the oldest democracies and civil societies, where, under the multiple banners of post-Modernism, the very autonomy of reason and conduct has come under suspicion. The ever more fashionable view is that the long prized autonomy of thinking and action is a delusion, quaintly obscuring how all reasoning and conduct is irreducibly caught within given frameworks that impose standards of truth and right relative to historical epoch and cultural milieu, as well as to particular standpoints within each historical-linguistic community rooted in class, gender, race, sexual orientation and the like. Far from being able to escape the hold of givenness and question reigning authority in theory and practice, reason and conduct are ineluctably foundation-ridden, condemned to lay claim to values that always issue from privileged terms that cannot be questioned precisely because all attempts at justification can only proceed upon their basis or some such other arbitrarily given paradigm.

In this way, the problem of foundationalism, so roundly denounced as a dilemma sabotaging the traditional aspirations of theoretical and practical philosophy, has now been wearily accepted as a fate from which there is no escape. The divide between foundationalists and anti-foundationalists has turned out to be no divide at all. For the self-proclaimed anti-foundationalists have simply abandoned challenging the hold of foundations upon reason and conduct, resigning themselves instead to observing the conversations of mankind, deconstructing philosophy with an edifying metareflection and metaethics, becoming self-conscious of the conceptual schemes that underlie all normative discourse, and at best engaging in a reflective equilibrium that points out how given views might be adjusted to conform more consistently with the given foundations they rest upon.

Many a latter-day anti-foundationalist would like to claim that such sober resignation is the only appropriate basis for a liberal society devoted to enabling all to pursue their vision of the good on an equal footing with others. Yet it is hard to see how the embrace of edification fits any less the new aristocracy of the *Übermensch* who openly imposes perspectival particular aims upon all others in a consistent exercise of value positing, which, after all, can be little else but a will to power if all values are arbitrary stipulations.

Is the yawning discrepancy between current theory and practice, between the post-modernist denial of autonomous reason and conduct and the pervasive march towards institutions of freedom, a harbinger of a counter-revolutionary upheaval, overturning the fundamental drift of modernity, or might it signify instead a theoretical blindness to an alternative philosophical strategy that overcomes foundationalism without abandoning the autonomy of reason and action?

Resolving this question requires reconsidering how foundations pose a problem for theory and practice that is neither fortuitous nor unavoidable. Although the endeavor of moving from opinion to knowledge is not entirely akin to that of moving from given convention to valid conduct, each involves parallel challenges of justification, eliciting common strategies in response. These approaches fall into three fundamental varieties, of which two encompass the entire foundationalist/anti-foundationalist spectrum of current fashion, and of which one remains the ignored odd man out, whose character and validity can best be understood by following out the breakdown of its two more familiar competitors. As it turns out, the status of freedom in theory and practice is determined by the very outcome of exploring each of these strategies in succession, first in the field of theoretical philosophy and then in the field of ethics.

The Appeal to Privileged Givens in Theoretical Philosophy

Whether it be in regard to thinking thinking or to thinking what is other to thought, philosophy has traditionally found it natural to base the validity of its claims in some privileged given, providing the foundation of ultimate justification for the conception of valid reasoning or of true reality. Any other option has appeared hopeless. For if the uncertainty, the particularity, the subjectivity, the conventionality, and the otherwise conditioned character of belief is to be overcome, how else can it be accomplished other than by discovering a given content that is immediate and thereby dependent on nothing else, that is unconditioned and thereby free of any relativizing factors, that is universal and not limited by particular circumstance, and that is in all these respects, presuppositionless. Insofar as truth appears to require presuppositionlessness, universal validity, and self-grounding, what else could command such authority other than some given content that is what it is by nature, rather than as a product or construct of some other factor? Moreover, such a content must not only be immediate, but serve as a first principle, providing a self-evident, self-grounded foundation from which all other truths are determined. If other contents cannot be shown to be grounded in such a privileged given, then how can they possibly rise above the uncertainty of opinion and share in the objectivity and unconditioned universality that such a first principle can alone command?

This applies as much to conceiving thought as to conceiving reality. If reason cannot be shown to rest upon one or more self-evident principles, from whom all other logical moves are derived, how can reasoning claim to have any validity? Analogously, if reality cannot be shown to rest upon a fundamental structure presupposing no other, given immediately to a passive contemplation of what is, how can philosophy pretend to offer a theory of reality that is not marred by a reliance upon unsubstantiated factors that are simply taken for granted?

Such are the considerations that have led what has been alternately labeled metaphysical, dogmatic, or precritical philosophy to appeal to privileged givens as the foundation of truth. Whether these privileged terms comprise first principles of reasoning or first principles of reality, they represent contents that are understood to be immediately given by nature rather than by convention, and thereby to be presuppositionless, unconditioned, universal and self-grounding. As such, they are apprehended by a contemplation that passively receives them, rather than constructs them, doing so not in virtue of any demonstration, but rather by direct intuition. Accordingly, this metaphysical approach characteristically begins its inquiry by asking what is and directly discovering the fundamental nature of things.

However, whether this investigation take the form of an ontology of being as such, or an ontotheology of the highest being on which all else depends, or regional ontologies of different spheres of existence, it cannot help but be plagued by a dilemma that equally afflicts any effort to establish the first principles of reason. If fundamental normativity resides in some privileged given, how can any content be certified to qualify as the sought after foundation? What can possibly decide which given content commands exclusive authority as that presuppositionless, unconditioned, universally valid term upon which all justification is rooted?

Clearly, to search for mediating reasons would only be self-defeating, for if the putative foundation owes its privileged role to something else, it forfeits its own constitutive primacy.

Can the first principle then be identified regressively, by certifying that all else, be it logical reasonings or all other beings, rests upon it? Yet, how, without reliance upon the foundation of justification, can one certify that all else rests upon any given content? On what basis can one determine what is the totality of reason or being, or for that matter, what constitutes a proper derivation of that plenitude from the putative foundation?

Appeal to an indirect proof is of little avail. Showing how all known given alternatives collapse through their own incoherence can hardly certify the absence of other unknown options. Moreover, even if one somehow could rule out every other alternative, that still would not legitimate the remaining candidate unless one appealed to a further principle, such as a law of excluded middle, whose own justification would not only remain in question, but whose supporting role would once again undermine the alleged primacy of the term it legitimates. And even if, following Aristotle in his argument for the principle of contradiction, one might show that no determinate opposing view could be advanced without endorsing the vaunted candidate, this would still rest upon prior assumptions about the conditions for something to be or be meant.

The Foundational Dilemmas of Teleological Ethics

Completely parallel difficulties apply in ethics, when the justification of conduct is sought in privileged givens. This is the basic strategy defining teleological ethics, which follows the all too plausible path of distinguishing between valid and invalid conduct according to whether it embodies the antecedently determined form of the good.

The rationale for measuring conduct by the good simply translates into another idiom the same line of reasoning that leads metaphysical philosophy to seek foundations for reality and reason. Since all conduct, being voluntary, aims at some end, how else can action be evaluated other than by determining a privileged goal, not only of inherent worth, but providing the rationale for the pursuit of all other ends? Only such a sovereign end-in-itself can provide an unequivocal standard for conduct that does not fall into either the infinite regress of instrumental action, for which no aim can be found whose value does not lie in serving as a means to something else, or the insoluble quandary of choosing among a plurality of ends-in-themselves. To avoid condemning action to an endless pursuit for some reason to choose one goal rather than another, a highest good must be discovered, whose validity lies nowhere else but in its own content, which, drawing its validity from no other source but its own immediate givenness, occupies the privileged role of first principle of conduct. Accordingly, ethics can be nothing other than a science of the highest good, which, given its sovereign character as the subordinating end of all other pursuits, is ultimately a form of political life. Conduct will therefore be valid not by exercising its own autonomy in determining its ends, but by embodying the given content that reason prescribes for this self-sufficient, unconditioned rule of living.

Yet, just as first principles of reason and reality are ready victims of sceptical challenge, so the highest good eludes any satisfactory determination. Defined simply as the sovereign end-in-itself to which all other pursuits are subordinated, the highest good is nothing but the formal scaffold of any prevailing political rule that succeeds in lording over all activity, reducing ethics to a nihilism of might makes right. This formality cannot be overcome by appealing to human nature and trying to discover a given function whose best fulfillment might give content to the highest good. Not only can the identity of a human function be questioned, but so can its prescriptive status. For if it is simply a fact of species being, how can it possibly have any independent normative significance? Even if one were to characterize the privileged human function as a matter of acting in accord with reason, this remains an utterly empty injunction, since all it does is place us once more before the problem of ethics, namely, of determining what indeed it is to act in a rationally justified manner.

The basic problem is that any attempt to legitimate a particular content as that of the highest good will require appeal to other factors. If these themselves are normative grounds, they undermine the supreme primacy of the highest good, whereas if they are devoid of normative significance, they cannot play any legitimate role in specifying what is of paramount intrinsic worth. Moreover, although the highest good, as that end-in-itself to which all other activity is subordinated, comprises a ruling activity that is for its own sake, it still serves as an antecedent form that conduct needs to embody in order to be valid. Yet how can conduct both embody something determined prior and apart from itself and be action for its own sake? Further, if the mere existence of a mode of conduct cannot supply it with normative validity and no given content can unequivocally qualify itself as the highest good, how can any goal be certified as a valid measure of conduct without appealing to the manner of its selection?

The Appeal to a Privileged Determiner

On all accounts, be it in theory or practice, the metaphysical strategy of rooting ultimate justification in a privileged given ends up revealing that the only reason for one content rather than another serving a foundational role is that it has been conferred that position by the theorist. No matter how the foundation be formulated, it is presented by a direct reference to what is that perennially takes for granted the adequacy of the knowing that claims primacy for its reputed givenness. Consequently, the failure of the appeal to privileged givens naturally shifts attention to the positing responsible for producing the putative foundation. Instead of relying upon a passive contemplation making direct reference to what is by nature, ultimate justification must be sought elsewhere, and where else but in that determiner capable of determining what counts as true and objective, not by offering privileged contents, but by engaging in a privileged act specifying the very domain of normative validity?

In this way, philosophy has made its transcendental turn away from any dogmatic direct reference to reality, choosing instead an indirect route that supplants the appeal to privileged givens with an appeal to a privileged determiner. In theoretical philosophy, this move entails a foundational epistemology that mandates as the first task of philosophy uncovering the structure of knowing or reference responsible for constructing what counts as true.

The embrace of a foundational epistemology as a preliminary for making possible objective knowledge might appear paradoxical at first glance. Since any direct reference to things in themselves must be ruled out, how can the turn to investigate knowing in its own right possibly establish the objectivity of knowing's claims? The very standard for adjudicating between competing beliefs seems to be wanting, for if all that can be legitimately considered is knowing and the truth of knowledge resides in correspondence to its object, how can that correspondence ever be certified by an investigation that restricts its gaze to cognitive structure?

Although the transcendental turn offers itself as the necessary preliminary to a valid conception of reality, the only way it can serve that role is if its conception of knowing ceases to be such a preliminary and instead contains within itself the determination of the object of knowledge to the extent that it can be known with more than pragmatic assurances.[1] For only if objectivity is determined by the structure of knowing can an investigation of knowing apart from reference to objects in themselves possibly bear upon the objectivity of knowledge.[2]

In this way, the transcendental investigation of knowing skirts the dilemma of the representational model of knowing. That model rendered knowledge unattainable by construing knowing and its object as given independently from one another, precluding any direct comparison of knowing's representations and their referent. When, by contrast, the object of knowing is taken to be a construct of knowing, an investigation of cognition can lay hold of what is knowable, since objectivity is precisely what knowing puts into its own representations. These achieve objective reference not through any comparison with things-in-themselves, but rather through relations among representations that can be shown to be necessary to the very structure of cognition, and in that sense, intersubjectively valid, rather than merely subjectively at hand.

Consequently, the transcendental turn does not simply alter the order of philosophical investigation by putting epistemology first, in place of ontology. Rather, transcendental philosophy equally transforms objectivity into a construct of the structure of cognition. What counts as valid is therefore no longer immediately given, but rather mediated by a structure of reference that is to be certified as the privileged epistemological foundation from which and for which objectivity is determined.

It is therefore no accident that objectivity gets construed by the great pioneer of the appeal to privileged determiners, Kant, as a domain of nature governed by the necessity of efficient causality. To begin with, since objectivity can now be legitimately known only insofar as it is determined from without by the structure of knowing, objectivity will be ordered by an external necessity. Moreover, since what is universal and necessary in objectivity issues from the unitary structure of knowing, which applies equally to all possible objects of knowledge, objectivity will have to be a nature governed by laws that apply to objects in general, irrespective of what kind of thing they are. Any qualitative differences between objects will be contingent givens of experience, whose sorting into "natural kinds" will depend upon corrigible observations of family resemblances devoid of any necessity or strict universality. If objects are to be known as enduring substances, they therefore will not be modeled after artifacts, combining an intelligible form imposed upon an indeterminate matter, but rather as entities whose lawfulness pertains to their homogeneous sensuous materiality, enabling them to be known by means of an empirical science.

This limitation poses notorious problems for the Kantian version of transcendental philosophy, for it excludes from the domain of objectivity living organisms, objects of beauty, and, above all, rational agents, none of which are intelligible solely in terms of efficient causality. With living selves unable to appear to themselves or to one another as such within the field of objectivity, self-consciousness, so central to the Kantian conception of cognition, becomes just as unfathomable as moral relations, which involve the mutual recognizability of persons.

These problems, together with Kant's equally notorious reference to things-in-themselves and his metaphysical deduction of the categories, might appear to be remedial failings of a first trailblazer, rather than fatal difficulties endemic to transcendental philosophy in general. And indeed, subsequent philosophy has been largely occupied with saving the transcendental project from the acknowledged blunders that Kant committed in pioneering its strategy.

Starting with Fichte and continuing with Husserl, latter day transcendental philosophers have recognized that Kant's residual reference to things-in-themselves involves just as much a retreat to a dogmatic appeal to the given as does Kant's appropriation of the table of categories from the forms of judgment of traditional logic. To remedy these inadvertent lapses, both have tried to eliminate any transcendent domain beyond the limits of transcendental constitution and to extend transcendental constitution beyond empirical objects to include the categories and forms of judgment themselves. After all, if validity cannot reside in privileged givens, then no knowledge claims of any sort can legitimately be made that cannot be referred back to the structure of cognition.

For similar reasons, thinkers from Schelling to Heidegger and Wittgenstein have attempted to remove the divide between the domain of objectivity and the transcendental structures themselves. Noumenal subjectivity has been replaced by such "objectified" conditions of knowing as the interest-laden being-in-the world of *Dasein*, and intersubjective structures of linguistic reference and cultural interpretation. Unless the conditions of knowing can be removed from the noumenal realm and brought down to earth in the form of tangible practices, such as language games and the like, how can the transcendental structures be known with any more authority than the objects of knowledge they are reputed to make possible?

Yet can either of these complementary extensions of transcendental argument surmount the difficulties at which they are aimed? If, on the one hand, all reference to the given is strictly forbidden, leaving an internal realism wherein objectivity is explicitly the stipulation of knowing, to what degree is solipsism avoidable? Can idealism be refuted without situating knowing in a framework that it has not totally constituted?

Moreover, can all features of the transcendental structure be determined in just the same way in which objects of knowledge are transcendentally constituted? The case of Husserl is instructive in this regard. Although Husserl, to his credit, is pointedly aware of the need to avoid any direct reference to the conditions of knowing, he recognizes that the moment an aspect of the transcendental structure is traced back to some act of transcendental constitution, that very act must itself be constituted by a further transcendental reflection if it is not to represent a new given to which dogmatic reference is made. Yet, as Husserl must admit, so long as object and act of knowing, *noema* and *noesis* are differentiated, as must be done to prevent intentionality from collapsing, transcendental investigation can *never* account for all its terms. Whenever an act of knowing is shown to constitute a type of object, that act stands in need of a further act to account for its own determination for knowing.

Phenomenological investigation must thus remain an infinite striving, whose incompleteness testifies to an ineradicable element of givenness in the transcendental structure itself.

This residue of givenness in transcendental investigation should come as no surprise. The moment that foundational epistemology supplants ontology as first philosophy, the transcendental investigator cannot help but make direct appeal to a certain structure as the condition of valid knowing. That the chosen structure is adequately described and that it has the exclusive privilege of determining objectivity are matters that cannot be established *after* the transcendental investigation has purportedly legitimated the scope and authority of knowledge, for the character of the conditions of knowing are, as such, the prior determinants of true knowledge, from which knowledge of particular objects can first be gained. It matters not whether the transcendental conditions be described as noumenal subjectivity, linguistic practice, hermeneutic reflection, relations of production or in any other terms. In every case, they occupy their privileged role as the constituting structure of objectivity insofar as they have their character and status antecedent to the knowledge they make possible. For this reason, they can only be referred to just as dogmatic philosophy made direct reference to its privileged givens. For although the investigation of the conditions of knowing is intended as a necessary preliminary to any knowledge of particular objects, it itself engages in a knowing of a subject matter of its own, exercising what Hegel would castigate as a "knowing before knowing".

Transforming the transcendental conditions into tangible practices having the same character as the objectivity they constitute cannot overcome the anomaly. Such a "naturalization" of epistemology only introduces the confusion of what Husserl branded as "psychologism", where the putative conditions of knowing are described in terms of the very objectivity that must be accounted for to begin with as something determined by the structure of knowing. When, for instance, a Quine analyzes the limits of reference in terms of the empirical behavior of speakers, the description of that behavior is immediately given as anything but ontologically relative through a direct reference whose legitimacy is no more questioned than Wittgenstein's own reference to language games.

The underlying problem bedeviling all of these efforts of transcendental philosophy is the abiding discrepancy between the knowing exercised by the transcendental investigator and the knowing under critique. Logically speaking, this discrepancy exhibits the perennial incoherence of any foundationalism: namely, that the privileged factor that confers normativity does not and indeed cannot itself enjoy the validity it provides. In the case of foundational epistemology, where the appeal to a privileged determiner has supplanted the appeal to privileged givens, the discrepancy consists in how objectivity is constituted by the transcendental structures of knowing, whereas these very structures, as prior determiners, have their character antecedent to their own act of constitution. As a consequence, they cannot possess a like objectivity of their own without canceling the distinction between the conditions of knowing and objects of knowledge, a distinction upon which the very possibility of a transcendental investigation rests.

The need to overcome this discrepancy is nevertheless inherent in the transcendental turn, once due attention is paid to the problem of removing any last residue of privileged givenness and the dogmatic reference to which it corresponds. Transcendental inquiry must call into question the authority of its own transcendental knowledge and certify the validity of its own knowing of the knowing under critique. In order for this transcendental knowledge to be valid, it must have precisely the same character that it discovers to be constitutive of valid knowing. Otherwise its cognition of the conditions of knowing will not conform to the standards of true knowing. Yet, for this to be the case, the metareflection of the transcendental philosopher must lose its difference from the knowing it reflects upon, or, in other words, knowing must do its own critique. Since what foundational epistemology knows is what is to be determined to be valid knowing, its own transcendental knowledge consists in a knowing of valid knowing. However, in order for this knowing to conform to what it knows, valid knowing must itself be a knowing of valid knowing. Then, however, the difference between knowing and its object disappears, and with it, any distinction between the conditions of knowing and what they condition. Transcendental inquiry thereby annuls itself, for knowing can be investigated on its own as a preliminary to obtaining knowledge of objects only if knowing and its object can be differentiated. Or, to put the outcome in purely logical terms, the conferring source of normative validity ceases to be a privileged determiner of something other than itself, thereby shedding its own given character as a prior condition.

The Dilemmas of Liberalism

The significance of this result is most palpable in the complementary form it takes when ethics abandons the search for a highest good and instead shifts the locus of normativity in conduct from the given content of ends and institutions to the source or procedure from which they issue. With the breakdown of teleological ethics exhibiting how no actions or conventions can enjoy validity based upon the content they embody, agents are liberated from any duty to realize any ends alleged to be given by nature or dictated by reason independently of their willing. By itself, this liberation from the external prescription of substantive ends does not confer legitimacy to the liberty of individuals to pursue ends of their own choosing. The inability to determine a highest good leaves equally open the option of nihilism, for which liberty has no more value than any candidate for the good. Nevertheless, just as the repudiation of the metaphysical search for first principles of reason and reality made it plausible to turn to investigate knowing as a preliminary to comprehending what is, so the denial of a determinable highest good paves the way for an appeal to the most plausible of privileged determiners in the field of practice, the individual will underlying all conduct and institutions. Instead of residing in conformity with the antecedent content of the good, normative validity now consists in being determined by whatever exercise of will is ascribed primacy. In this respect, ethics becomes a procedural theory, where valid practice enjoys authority not because its content conforms to any given standard, but because it can be shown to be determined by a privileged method of willing. The latter plays the archimedean role of being bound by no prior norms, but instead automatically conferring legitimacy to whatever issues from it.

This foundational exercise can be interpreted in an empiricist manner, according to which liberty consists in the entitled freedom to act in pursuit of given desires. Due to the repudiation of any prior rational hierarchy of ends or corresponding virtues and character, these desires all stand on just as much a par as each and every will. Consequently, the primacy of the liberty of desire translates into a utilitarian principle by which normative validity lies in any course of action or institutional arrangement that promotes the greatest aggregate satisfaction of the desires of all.

However, the empiricist interpretation of liberty thereby binds the will to a pursuit of given ends, whose contingent facticity makes any utilitarian calculus problematic and equally reintroduces a domain of privileged givenness. Consequently procedural ethics is impelled to overcome such heteronomy and interpret its privileged exercise of freedom in an a priori fashion, where entitled freedom consists in acting independently of given desires according to a rational form of willing whose privileged exercise comprises a principle of right preempting any prior interpretation of the good.

Either way, the liberty that is accorded primacy functions as a prior principle of valid conduct and institutions. They are legitimate by being determined by it. Consequently, whether the privileged procedural factor be construed as a liberty of desire or the autonomy of a practical reason, it has a given character that precedes what its application determines to be moral and just. As a result, neither itself nor the conduct and institutions that issue from it exhibit self-determination proper. The practice that owes its normativity to issuing from a privileged procedure is determined by that prior procedure, which, in determining that derivative domain, does not give itself any further character, but rather specifies something other than itself. Logically speaking, the freedom embraced by these varieties of liberalism exhibits the two-tiered structure of a logic of essence, where one term posits another. By contrast, in self-determination, what does the determining is one and the same as what is determined. When the will is treated as a principle, it is relegated to the status of a prior positor whose character is given antecedently to its act of positing. In other words, just as the transcendental conditions of knowing could not help but possess an element of givenness in their own prior description, so procedural ethics offers a principle of willing whose own character is just as given as any candidate for the highest good.

This residual element of givenness afflicting the appeal to liberty shows its problematic character on two fronts: first, in face of the specter of formalism haunting its efforts to unequivocally derive determinate norms and institutions from its privileged procedure, and then, secondly, and most crucially, when the very authority of the proffered principle of willing is called into question.

Formalism enters in because the privileged procedure comprises a mode of willing whose character is given prior to its own exercise. Accordingly, it constitutes a capacity, rather than an actuality, which must draw the content of its actual exercise from some other source. Thus, the privileged liberty of utilitarianism draws its ends from empirical desire, whereas the formal willing of Kantian autonomy must still rely upon the given content of maxims to supply it with anything to test for universalizability. Yet, under these conditions, how can the determining of the privileged procedure really signify a departure from the teleological reliance upon privileged givens? In order to determine whether a course of action conforms to the formal principle, must not some additional criteria of application be introduced, undermining the exclusive normativity of the procedure? In this vein, for instance, Hegel points out that Kant's disqualification of borrowing funds with no intention of repayment as non-universalizable illicitly presupposes a commitment to property.

Habermas' communicative ethics brings this dilemma to perhaps its most absurd and telling form when he proudly admits that his ideal speech situation of non-distorted communication is such that one cannot predict, prior to its exercise, what norms and institutions will be agreed upon as sanctioned practices. Yet since his privileged speech situation is not an actual normative institution, but an ideal procedure for determining what is just, there can never be any opportunity to observe what might issue from its operation. Hence, communicative ethics collapses into sheer indeterminacy, effacing itself as an alternative to the teleological option against which it rebels.

Even more fatal, however, is the dilemma inescapably surrounding the legitimacy of the privileged procedure, given its own repudiation of rooting normativity in any given end or institution. Whereas the privileged procedure should not be bound by any independently given norms, everything of normative standing should issue from its application. Yet how then can its own content be justified? Does not, for instance, Rawls' description of the original position or Habermas' characterization of the ideal speech situation rest upon a prior commitment to liberty and a corresponding impartial respect for individual interests, which, as such, cannot issue from the procedure it grounds? Or more generally, must not any privileged procedure owe its own determination to a source given independently of its own application, thereby undercutting the archimedean role it otherwise claims for itself? This is the very same problem that afflicts the givenness of the transcendental conditions, whose own character cannot possess the certified objectivity they themselves reputedly constitute.

Little relief is obtained by openly admitting the foundational character of the privileged procedure, as Rawls effectively does in acknowledging how the original position merely conforms to our given moral intuitions and thus can only frame *a*, rather than, *the* theory of justice, bringing a degree of consistency to the historically relative ethics of our community. This admission can only bear upon the fate of ethics if a supplementary argument shows that no other alternatives are possible. Yet can such an argument be delivered by a thinking resigned to the foundation-ridden relativism of reflective equilibrium?

As in the case of foundational epistemology, the dilemma demanding a solution lies in the abiding discrepancy between what possesses normativity and what confers normativity. The true can no more be determined by what lacks truth than right can be determined by what stands outside the domain of valid conduct. In each case, the problem of foundationalism points towards eliminating the divide between foundation and what is founded, between procedure and derivative norms; yet canceling the divide is precisely what eliminates the very enterprise of these respective theoretical strategies.

The collapse of procedural ethics thus confirms how in practical as well as in theoretical philosophy the appeal to privileged determiners fares no better than the appeal to privileged givens. With the demise of these complementary foundationalisms, normative validity may be freed from the dilemmas of conformity with any privileged givens or determination by any privileged determiner. Yet does this liberation signify an end to philosophy, where the collapse of the dual foundationalisms of foundational ontology and teleological ethics on the one hand, and foundational epistemology and liberalism on the other, leave no other option in their wake? These two brands of foundationalism may well garner the lion's share of philosophical history, but can this fact legitimate the post-modern turn to supplant philosophical investigation with edifying deconstructions of given theories?

The Dogmatism of the Post-Modern Turn

In order for the failure of the two varieties of foundationalism to condemn philosophy to impotence, a valid argument must be made to establish that they exhaust the possibilities of philosophical inquiry. But can the would-be grave-diggers of philosophy succeed at their therapeutic game if, to do so, they must make definitive claims about the options of philosophical reason? Can philosophical argument be employed to cure us of philosophy?

Far from overcoming philosophy, post-modernism's turn to edifying deconstruction is, if anything, a regression to the very foundationalism it seeks to avoid. For when the post-modernist depicts the conversation of mankind as foundation-ridden and then proceeds to lay bare the conceptual schemes supposedly underlying various philosophical projects, the post-modernist reverts to dogmatic foundationalism by making direct reference to the foundation-ridden practice of discourse and then treats it as a transcendental condition of knowing, never bothering to reflect on what legitimacy these claims can have in their own right.

If post-modernism denies that it mirrors the reality of discourse or transcendentally constitutes historically contingent paradigms of knowledge and conduct, it fares no better in opting for a holism in which truth and right are seen to reside in coherence with our historically given web of belief and the practices that underlie and sustain it. Once again, the post-modernist must explain how reference can be made to that encompassing web of theory and practice in terms of coherence, since that web itself either cannot be known at all, given the lack of anything further with which it can cohere, or it must be known directly, in flagrant contradiction of the coherence theory itself.

The Normativity of Autonomy in Theory

How then can philosophical investigation possibly proceed given post-modernism's failure to condemn philosophy to foundationalism and given foundationalism's own collapse in both its forms? By themselves, the breakdowns of the appeals to privileged givens and privileged determiners in theory and practice offer little more than a negative prescription: the quests for truth and right must free themselves of foundations and somehow operate without making any assumptions concerning either the content of their respective topics or the method of their treatment. With respect to philosophy in general, this rejection of foundations mandates that philosophical investigation must proceed without any preconception concerning either form or content. Indeed, since all questions of the subject matter and method of philosophy are themselves philosophical issues they must be addressed within philosophy itself, rather than accepted as matters settled prior to the exercise of philosophical reason. Hence, escaping the appeal to privileged givens and privileged determiners is tantamount to achieving a thoroughly radical presuppositionlessness, not in the bogus way of stipulating specific contents and procedures as if they were beyond assumption, but by avoiding beginning with any determinate givens whatsoever. Yet, if this negative prescription leaves philosophy with the prospect of beginning with nothing at all, it has a positive significance that is the other side of the negative freedom from foundations.

Whereas philosophy can begin with no determinate claims concerning reality or knowing, any determinate subject matter that arises for philosophical inquiry must be generated wholly within that inquiry and arise in an ordering that itself emerges without reliance upon any externally given principles or procedures. In this respect, a philosophy free of foundations is, in the first instance, a theory of determinacy as such. Neither metaphysical or transcendental inquiry can provide such a theory precisely because any such foundationalisms proceed from given determinacies of one sort or another, precluding any account of determinacy in general. Only an investigation beginning without any determinate foundations, without any determinate claims about what is or about knowing, can possibly account for this most minimal of contents, whose lack of treatment testifies to the dogmatism of any other approach. Yet if philosophy can be and cannot help but be a theory of determinacy by starting without foundations, its account of determinacy is no less an account of self-determined determinacy, that is, of self-determination per se.

Philosophy without foundations turns out to present self-determined determinacy because, given the exclusion of any external factors, whatever content and ordering it can possibly arrive at will have to be self-generated. Starting with indeterminacy, philosophical investigation can only consist in a self-developing conceptual process, whose subject matter orders itself. Since this self-development proceeds from no given substrate or determining standpoint, it will not comprise the self-determination of any given content, but rather will be self-determination per se.

This does not signal the embrace of a new foundation, where self-determination supplants given determinacy and determined determinacy as the ground of normativity. Precisely because self-determination is what it is as a result of its own determination, it can have no given content. For this reason, self-determination must proceed from indeterminacy and any attempt to treat self-determination as a principle automatically falsifies it, transforming it into a privileged determiner, whose own content is given rather than self-informed.

By issuing in the self-exposition of self-determination as the positive side of the liberation from foundations, philosophy is able to attain the self-grounding, unconditioned universality and presuppositionlessness that has always eluded philosophical reason as long as it appealed to privileged contents and privileged standpoints. Self-determination is the very logic of self-grounding precisely because it owes its character to nothing but itself. By the same token, self-determination enjoys unconditioned universality insofar as what it is is independent of any condition or particular factor. And finally, self-determination is the positive outcome of presuppositionlessness precisely because it alone provides a content that rests on no prior terms.

In all these converging respects, self-determination turns out to be the very logic of normativity. The truth of the categories comprising its self-development lies not in correspondence to any independent givens, nor in coherence with any antecedent totality, nor in being constituted by any privileged conditions of reference. It instead lies in the very freedom from extraneous determination, which positively expressed, could be called categorial immanence. That is, the truth of the categories of a foundation-free philosophy will reside in how they arise from one another in an order wedded to nothing but their own content, proceeding from indeterminacy. Since self-determined determinacy comes to be what it is only as the outcome of its process, what its constitutive categories are categories of is only determinable as the result of the process. By the same token, the ordering principle or method of their development will also only be at hand at the very end, for the ordering unity of the development is nothing other than self-determination itself. Consequently, completely contrary to foundationalism, the subject matter and method of philosophy will not be given at the outset, but be arrived at as the outcome of philosophical investigation. In this way, philosophy will exhibit a radically autonomous reason, whose negative and positive freedom from any external determination is the very mark of its normative validity.

Philosophy will thereby be systematic to the very degree that it is anti-foundationalist. Whereas foundational theories root the introduction and ordering of their content on arbitrary givens, philosophy without foundations will have to present the self-development of its subject matter, whose own resultant determination will confirm in precisely what order it must present itself. At each stage along the way, no term can be legitimately introduced that involves more than those that are already at hand. Yet, because the whole process proceeds from indeterminacy, rather than from first principles, each prior term will have to give rise to what is other than itself as the outcome of its very own determination. In this way, each move will operate as a determinate negation, where the only resource providing for the introduction of new content will be the preceding categories, whose own specification will involve their own supersession and passage into a term incorporating them without being reducible to them. Accordingly, at every point along the way, what gets treated will be just as non-arbitrary as when it comes into view. Although this strict systematicity stands so opposed to the familiar paths of metaphysical reference and transcendental constitution, it is nothing more than the fulfillment of the perennial demand of philosophical self-responsibility: that contents not be introduced until after all their components have been established.

This identification between freedom from foundations and systematicity, whereby philosophy achieves presuppositionlessness by proceeding as a theory of determinacy presenting the self-exposition of self-determination, can further be identified as the science of logic.[3] As the prescriptive discipline that seeks to think validly what is valid thinking, logic can hardly achieve its constitutive aim if it allows any aspect of its method or content to be preconceived. Whereas other disciplines have a subject matter distinct from the method by which it is examined, prescriptive logic investigates precisely what it must employ to capture its prey. To be the valid thinking of valid thinking, prescriptive logic's form and content must be one. For this very reason, logic cannot begin with any preestablished method without utterly begging the question. By the same token, logic cannot begin with any determinate subject matter, since what valid thinking is is precisely what its own investigation must establish. Consequently, prescriptive logic must begin without any foundations and end up presenting the self-thinking of valid thought, which, as such, cannot help but be a wholly autonomous development. Since its autonomy proceeds from no antecedent content or method, prescriptive logic ends up being the same theory of self-determination per se that philosophy begins with once it frees itself from foundations.

These brief anticipations of how reason attains normative validity by achieving a foundation-free autonomy are so contrary to prevailing fashion that they have an almost unintelligible ring, despite the detailed efforts long made to bring them to fruition, above all by Hegel in his much neglected *Science of Logic*. The normativity of autonomous reason, however, has a tangible corroboration of sorts in the completely analogous overcoming of foundations in ethics and in modernity's parallel struggle to replace tradition with institutions of freedom in every sphere of conduct.

The Normativity of Autonomy in Practice

Although liberalism already brings freedom to center stage in ethics, it commits the foundationalist blunder of treating self-determination as a first principle, transforming it into a liberty defined by nature, by the self, or by some other factor prior to the practices and institutions that are ascribed legitimacy by being determined by its antecedent form. As the dilemmas of procedural ethics make most patent, normativity in conduct cannot be coherently determined unless what confers normativity no longer stands apart from what possesses normative validity. In order for conduct to escape the insoluble dilemmas of foundationalism, valid conduct and institutions must thus be determined by nothing other than themselves. That is, normative validity in conduct can be none other than the reality of self-determination.

However, if freedom is not to revert to a determining principle of derivative institutions, self-determination can no longer be conceived apart from legitimate institutions. Instead, self-determination must itself comprise a system of institutions of freedom exhausting normativity in practice. The unity of this system must itself be the product of one of its own constitutive modes of self-determination if it is not to rest upon a given foundation undermining its autonomy. Only in this way can agency be genuinely self-determined, for only within a self-ordered system of enacted modes of freedom can agents engage in action where they determine both the form and content of their agency. So long as agency be confined to the isolated deeds of the single self, it cannot help but be defined in terms of the logic of positing, as a determiner whose own structure precedes what it determines. Admittedly, the corresponding "natural" will of the individual cannot be done without, since no institutions can function without their participants each employing a will whose faculty of choice is given prior to any of its decisions. Nevertheless, this faculty of choice no more determines which of the actions it can select count as valid practice than such conditions of discourse as language, consciousness, or, for that matter, digestion, can determine which of the claims they make possible is true. Only within the enacted contexts of a system of property relations, a moral community, a free family, a civil society, and a constitutional democracy, can individuals achieve genuine self-determination, determining both who they are and what they will in their "artificial" capacities as autonomous owners, moral subjects, family members, economic agents, and citizens. In so doing, they exercise objectively recognized modes of autonomy that cannot be performed apart from the interactions to which they belong, interactions which themselves consist in nothing but the respected exercise of the freedoms they make possible. Moreover, to the extent that one of these structures of interaction involves a political freedom of constitutional self-government that unifies the entire system of freedoms into a self-ordered whole, containing its standard within itself, individuals are able to engage in a self-determination that thoroughly grounds all dimensions of their autonomy. In this way, autonomous conduct animates a conventional world of its own whose normative character is determined independently of any given foundations.

Accordingly, just as philosophy in general overcomes foundations by systematically developing autonomous reason, ethics surmounts the challenge of post-modernism by conceiving the self-ordered system of the institutions of freedom. Insofar as this system consists in the respected reality of self-determination, justice is not a matter of embodying a highest good or being the outcome of the proper choice procedure or cohering with any historically given community. Valid practice instead consists in nothing but the concordant exercise of rights in all their different spheres. Therefore, what gives modernity its abiding standard can be none other than the philosophy of right, whose conception of self-determination prescribes what ought to be done.

Hence in theory or practice, to ignore the normativity of autonomy is to live on one's knees.

*****

The preceding argument for identifying autonomy and normativity and thereby overcoming foundations has been completely ignored by almost all contemporary theorists. This is due partly to their ingrained embrace of transcendental argument and partly to their inability to conceive of any positive alternative that does not revert to precritical dogmatism. These complementary tendencies are underscored by pervasive misinterpretations of Hegel, the great historical pioneer of doing philosophy without foundations. Chapter II seeks to remedy this situation by exposing how the prevailing schools of contemporary philosophy remain trapped within the dilemmas of transcendental argument and showing how Hegel's Phenomenology of Spirit *and* Science of Logic *trailblaze a workable strategy for achieving the autonomy of reason fulfilling the emancipation from foundations.*

Notes

1 Although, taking the example of Kant, one might be tempted to argue that the knowledge in question is only a priori objective knowledge and that empirical knowledge can be obtained by relying upon the individual matter of experience, knowledge of that material side of experience remains subject to irremovable doubts that can only be bracketed out by pragmatic agreement to accept a certain level of trust in observation, whose results must still always remain corrigible beliefs.

2 So it is that Kant, who offers his three critiques as preliminaries to a further metaphysics, ends up determining what is essential to nature, ethics and beauty in these critiques, adding only analytic extrapolations upon their doctrines in his subsequent *Metaphysics of Nature* and *Metaphysics of Ethics*.

3 For this reason, Hegel identifies the first part of his foundation-free systematic philosophy as *The Science of Logic*.

Two

Hegel's Remedy for the Impasse of Contemporary Philosophy

The Impasse of Contemporary Philosophy

In recent years there has been a growing convergence between the two mainstreams of contemporary philosophy, the so-called analytic and continental schools of thought.[1] Despite their divergent styles, they have reached common diagnoses of past philosophical problems, proposed common cures, and finally begun to acknowledge the kindred character of their respective enterprises. The emerging dialogue, however, has revealed not that truth resides in consensus, but that philosophy today has reached a common impasse.

Be it analytic or continental, contemporary philosophical inquiry has sought to surmount the dilemmas of traditional metaphysics and transcendental thought with two complementary projects that cannot help but fall victim to the very problem they seek to avoid. These corollary approaches are represented in the analytic tradition by ideal and ordinary language philosophies, just as they are represented in continental circles, on one pole, by structuralism and the pragmatic semiotics of Apel and Habermas, and, on the other, by hermeneutic philosophy. In each case, the chosen strategy follows from the awareness that philosophy can neither make unmediated truth claims about reality, directly describing the given, as traditional metaphysics had attempted, nor begin by characterizing some transcendental subject through which the limits of true knowing are established.

On the one hand, philosophy cannot directly ask, "What is?", due to the problematic character of any immediate reference to reality. Such reference claims truth for a content whose givenness is alleged. However, the presumed givenness of this content precludes appeal to any higher principle to adjudicate between it and any other content for which immediate being is claimed. Recourse to any such standard would introduce a mediating factor undermining the putative immediate givenness of any term to which it is applied. As a consequence, each competing content can have nothing supporting it but the claim that it is, an assumption as groundless as any other arbitrary assertion. Owing to its putative immediacy, each determinate truth claim thus can only be but a mere stipulation. This dilemma cannot be surmounted by attempting to show that some given content operates as a first principle of reality from which all else is derived and ordered. Any such attempt only reproduces the same problem on another level by leaving the content of the putative first principle itself just as immediately given as whatever standards of derivation and completeness that are employed to certify its grounding role. Once again, the metaphysical appeal to the given remains insusceptible of any justification.

Recognizing these metaphysical problems, contemporary analytic and continental philosophy has acknowledged the necessity of foregoing all immediate reference to reality and has chosen instead to investigate how truth claims are made. In making this turn to consider not reality, but the conditions of reference to reality, both schools have recognized the problem of doing so in the manner of Kantian transcendental philosophy, which appears to make immediate reference of its own to both the conditions of experience and the conditions of the object of experience. On the one hand, Kant is taken to task for metaphysically stipulating the character of the transcendental structure by conceiving it as a noumenal self determined through such unfounded devices as a metaphysical deduction of the categories, which simply adopts, with certain unargued modifications, the typology of judgment of received tradition. On the other hand, Kant is equally criticized for determining the object of experience with respect to a thing-in-itself, which is not transcendentally constituted, but metaphysically referred to as something immediately given *in res*.

To avoid these lapses into unmediated metaphysical reference, the two contemporary schools have attempted to conceive the conditions of making truth claims without referring to either any thing-in-itself or any acts of a transcendental subject. Instead of construing the object of knowing as the appearance of something outside knowing from which intuitions are received, they have taken the object of knowing to be something completely constituted by and within the very structure of referring itself. On the other hand, instead of conceiving any transcendental structure as a noumenal self hidden from knowing as much as any thing-in-itself, they have taken the practice of referring to be that in terms of which all reference is to be understood. Of course, this practice can be considered either to have an intrinsic universal character or else to be overdetermined, taking shape according to the particular context in which it occurs. Consequently, the process of referring in which all truth claims are constituted here gets alternately conceived as either an ideal structure of speech or as the given usage of a natural language.

On this basis the analytic tradition has made its linguistic turn and pursued the corollary strategies of ideal and ordinary language philosophy, wherein all questions of truth are reduced to questions of how truth claims are determined through linguistic practice, be it overdetermined or not.

For its part, recent continental philosophy has followed an analogous path. On the one hand, it has developed its own versions of ideal language philosophy, formulating it both as a theory of communicative competence that specifies the ideal speech situation under which legitimate discourse is possible and as a universal structuralism that uncovers the hidden forms of signifying by which meaning gets constituted. On the other hand, the continental tradition has offered its own analogue for ordinary language philosophy, hermeneutic philosophy. Under its banner, the irreducible condition of true knowing has been construed to be the contextually bound situation of interpretation, wherein discourse is predetermined by the given system of reference in which it operates. However this system is specified, the resulting historicity of knowing offers the same overdetermined transcendental framework presented by analytic ordinary language philosophy.

On the face of it, both versions of the convergent analytic and continental approaches seem to escape the particular difficulties of the thing-in-itself and the noumenal self. By making the practice of discourse what constitutes both the object of knowing and the knowledge of that object, they avoid any immediate reference to either subjective or objective reality.

Nevertheless, in so doing, they have hardly removed the central dilemma of transcendental argument, which is by no means a special affliction of the Kantian formulation, but concerns the status of the entire transcendental inquiry itself. In a word, what the two contemporary schools have failed to resolve is the problem of legitimating the discourse they themselves exercise in asserting the primacy of their chosen systems of reference. It matters not whether their system of reference be specified as an ideal or ordinary linguistic practice, or as semiotic structure, communicative competence, or the hermeneutic situation. Whatever its guise, the constitutive structure of reference remains a metaphysical stipulation so long as the discourse specifying it is not itself already constituted in terms of that structure.

The dilemma is simple enough. If the indicated framework of referring be the condition of all truth claims, then the philosopher's own characterization of that framework can claim no truth unless a further condition be met. Namely, this thematization by the philosopher must proceed according to the same critically established conditions of the referring it investigates. Otherwise, the authority of the transcendental investigation is itself uncritically assumed. To remove this residual dogmatism, the "metalanguage" of the philosopher must lose its metalinguistic transcendence and coincide with the discourse whose constitutive structures are being uncovered. To avoid any metaphysical reference to the transcendental structure itself, the philosophical practice of the ideal or ordinary language philosopher must thus become fully self-referential, which means that the truth claiming under consideration must perform its own critique.

What leaves analytic and continental thought at a common impasse is that the required equalization of transcendental argument with its object actually eliminates the very framework for doing transcendental philosophy of any sort, regardless of whether it makes the conditions of knowing a noumenal self, an ideal speech situation or an overdetermined hermeneutic context. This becomes manifest once one observes what happens when the discourse of transcendental inquiry becomes self-critical, forsaking all immediate metaphysical reference by becoming one and the same as the structure of referring under investigation.

To begin with, what transcendental discourse itself generically performs is a knowing of true knowing in terms of the conditions that make truth claims possible and give them their proper limit. The controversy between the different proponents of transcendental philosophy does not concern this general task, but rather the specific content assigned to knowing and the constitutive structure of its referring. Consequently, whatever its particular shape, if transcendental discourse is to exercise true knowing instead of an unfounded metaphysical stipulation, then it must relate to its subject matter just as true knowing relates to its object. Since transcendental discourse comprises a knowing of true knowing, it can validly embody the structure of true knowing, its object, only if the transcendental investigation, the knowing of true knowing, is the same as what it knows. For this to be true, true knowing must itself be a knowing of true knowing.

The achievement of this, however, removes not only the distinction between transcendental discourse and the knowing under critique, but also the distinction between knowing and its object, or between referring and its referent. Namely, if true knowing is itself a knowing of true knowing, then what knowing refers to is identical to knowing's relation to its object.

What makes this outcome of fatal consequence is that transcendental discourse can only be undertaken if knowing can be differentiated from its particular object. Only then can knowing be considered apart, independently of any specific knowledge, that is, independently of any objective reference. Yet if knowing is indistinguishable from what it knows, then, to use Kant's terminology, knowing cannot be examined prior to experience. As a result, when transcendental philosophy makes itself self-referential, eliminating all distinction between itself and the knowing it investigates, the accompanying equalization of knowing and its object eliminates the very possibility of transcendental discourse itself. By being driven to this result, transcendental inquiry testifies to the bankruptcy of its own enterprise. Since the analytic and continental schools have held on to its program, while merely substituting linguistic practice for noumenal subjectivity, the self-elimination of transcendental argument signals their common failure.

The Alternative of a Non-Transcendental Phenomenology

If this outcome indicates that philosophy can begin no more with any reference to knowing than with any reference to reality, it does not leave thought bound to the impasse of contemporary philosophy. In fact, the philosophical tradition has already offered an alternative to metaphysics and transcendental argument, an alternative that presents an all too neglected strategy for overcoming their difficulties and pursuing a systematic philosophy of an entirely different order. The original proponent of this alternative is Hegel and in the Introduction to his *Phenomenology of Spirit* he sketches out its basic strategy. In face of the dilemmas of contemporary thought, this strategy warrants reconsideration now more than ever, irrespectively of whether Hegel actually succeeded in carrying it out.

Needless to say, Hegel has commonly been interpreted as the final representative of the metaphysical tradition, who makes the last grandiose attempt to reach absolute knowledge of things as they are in themselves with a theory of subject-object identity.[2] Nevertheless, his approach is actually no such relic of the past, but as contemporary as can be.

Contrary to received opinion, Hegel begins by considering precisely that impasse at which today's thought has arrived. Confronting the failure of metaphysical and transcendental philosophies, Hegel asks how philosophy can begin at all. Their examples have shown that philosophy cannot begin with any immediate truth claims about either reality or knowing. This seems to leave one option open: that one begin with no given content whatsoever by casting aside all assumptions and resolving to think independently of any unmediated references to reality or any transcendental structures. Hegel recognizes, however, that such a resolve could not help but be a mere subjective postulate if philosophy began immediately with it.[3] Doing so would tacitly presuppose both the primacy of contentless indeterminacy and the primacy of that non-metaphysical, non-transcendental knowing that here presumably begins without any specific knowledge. In effect, this immediate resolve to think without assumption would assume prior to philosophical investigation that philosophy is properly presuppositionless knowing.

To avoid this recursion to metaphysical truth claims, Hegel offers the radical alternative of a non-transcendental phenomenology. It is conceived as an explicitly *positive* science observing what occurs to the project of foundational knowing common to metaphysics and transcendental argument when foundational cognition tests its own fundamental claims. What is at stake is whether this investigation, where the knowing under view does its own critique, can result in some threshold where knowledge claims can be made free from the pitfalls of metaphysical reference and transcendental constitution. If phenomenology can arrive at such a result, then, independently of all subjective resolve, a starting point will lie secured for a new type of philosophy that takes nothing for granted.

In line with this strategy, Hegel gives phenomenology its specific method and subject matter in direct challenge to the basic problem of metaphysics and transcendental philosophy. Their inquiries have shown how no immediate truth can be legitimately claimed by any knowing whose object or knowledge has some definite predetermined content. What Hegel therefore proposes is that, instead of making truth claims, one begin by stipulating knowing that claims truth for its knowledge by appeal to some given, and then observe how this explicitly presupposed subject matter develops itself by making and testing truth claims of its own. In this way, a wholly immanent critique can be undertaken of the strategy of knowing that justifies its claims by appeal to some foundation, be it construed as some factor *in res* or a transcendental condition.

In these terms, Hegel presents a phenomenological inquiry that is not only non-metaphysical in the traditional sense, but radically non-transcendental as well. Since this phenomenology will simply observe a structure of foundational knowing that it openly takes for granted as a given content, its investigation avoids metaphysics by making no claims concerning either the unqualified reality of its subject matter or the truth of the claims made by the subject matter itself. Unlike so many of his subsequent interpreters, Hegel is well aware of the problem that would arise if phenomenology did make such assertions, claiming either that it presents the true doctrine of knowing as it is in itself or that the truth claims made by its subject matter were those generic to knowing per se. If phenomenology followed that course, it would be but another version of transcendental philosophy, making the indefensible metaphysical assumption that it was itself rigorous science, laying bare the true underlying structure of all discourse. This is the fate of Husserlian phenomenology, which remains transcendental by claiming presuppositionlessness for its own observation, dogmatically assuming that all knowing must have the shape of intentionality as it is stipulated by Husserl.

By contrast, the phenomenology that Hegel here proposes forgoes all such unqualified assertion by openly accepting the limits of positive science. Stipulating the content it observes, this phenomenology appropriately admits that what claims do emerge are not truths in themselves, definitive of either reality or knowing. The observed truth claims are rather only beliefs generated by the subject matter, a subject matter that is itself ascribed no ontological or transcendental status, but merely taken for granted as a version of knowing posited by the phenomenologist herself.

Nevertheless, non-transcendental phenomenology is a very special positive science because its given object has the unique character of making its own truth claims and also testing them by itself. In so doing, the structure of knowing stipulated by phenomenology gives itself successive shapes of knowing, each with a different knowledge claim and a different standard of truth. Thereby the subject matter determines its own development, unlike in other positive sciences where the act of the investigator must be relied upon to introduce every new content. Consequently, the method of the positive science of phenomenology has the peculiar character of being what Hegel aptly calls a pure observation,[4] pure in that the phenomenological investigator need not interfere with the self-examination of the subject matter. For this reason, phenomenology has a singularly non-arbitrary character, even though it is only a positive science. Despite its stipulated subject matter, phenomenological discourse is ruled by the internal necessity that whatever content comes into view does so not by any intervention of the phenomenologist, but by being generated from nothing but the bare structure of knowing taken up at the start.

In the Introduction to the *Phenomenology of Spirit* Hegel indicates how these generic features all follow from the character of the given subject matter which the positive science of phenomenology begins observing. Hegel calls this presupposed content the structure of consciousness.[5] Although he will later attempt to confirm in his *Philosophy of Spirit* that consciousness is defined by the representational model of knowing that phenomenology addresses as a posit,[6] the introduction of conscious knowing here involves no further claim than that it denotes the cognitive structure that phenomenology stipulates for itself. The structure of consciousness thus denotes simply knowing that claims truth for its knowledge, referring its putative cognition to some given as the standard of its validity. Such a structure provides the appropriate subject matter if one is to forgo making all immediate truth claims and instead observe a given structure that makes them on its own. As accordingly stipulated by phenomenology, this structure consists in knowing that refers to what it knows as something both in relation to it and determined in itself independently of that relation.

These two aspects, which Hegel terms the being-for-consciousness and the being-in-itself of the object,[7] are necessary if the knowing under consideration is to make truth claims. Only with their distinction from one another can knowing have knowledge of what is not merely a subjective posit but putatively determined in its own right. Consequently, the knowing that phenomenology observes has this dual structure where, on the one hand, knowing's relation to its object is its knowledge, whereas, on the other hand, the truth of that knowledge is the known object taken by itself as that to which the knowledge refers.

Granted this characterization, it still might appear a contradiction in terms for knowing to refer to what is not merely in relation to it, but in itself independent of that relation. This problem poses no difficulty at all, however, just given the structure of the knowing under observation. Since this truth-claiming knowing consists in the polar relation of knowledge and truth, of what is *for* it and what is *to* it, what is in itself actually falls within knowing as one of its constitutive contrast terms.

For this important reason, as Hegel duly notes,[8] testing the truth of knowing's knowledge requires no introduction of any criterion of validity by the phenomenologist. If that were necessary, phenomenology would end up having to make metaphysical claims concerning what is the criterion of truth. This reversion to metaphysics need not arise precisely because the given structure of knowing not only claims truth for its knowledge, but contains within itself the standard by which its knowledge can be verified. This criterion of truth is none other than the constituent pole of in-itselfness, which is to knowing as that to which its knowledge should correspond. Since knowledge is knowing's relation to this content, the standard of truth for this knowledge is concomitantly given for knowing as its referent.

If this removes the traditional metaphysical problem of providing a criterion of truth, which is tantamount to gaining access to what is in itself, it also removes the transcendental problem of determining true knowing, which is the object of transcendental knowledge. That dilemma falls away as well, since phenomenology is equally relieved of having to apply the criterion of truth and thereby uphold some specific principle of method. As Hegel observes,[9] this difficulty is also overcome because knowing claims truth for its knowledge only by referring its knowledge to what it knows. In making truth claims, knowing considers not just what it knows, but *both* what it refers to and what its own knowledge is. Since knowledge and its referent are accordingly linked together for knowing as the corresponding terms of its own relation, knowing not only supplies the truth criterion of its knowledge. It further compares its knowledge with this its referent and only recognizes the truth of its cognition through this comparison.

Nevertheless, given the stipulated structure of knowing, the comparison of the standard of truth and knowledge cannot sustain certainty. Once knowing has its two constituents before it, relating them one to another, its truth criterion is not in itself any longer, as something given independently of cognition, but something for knowing, determined in virtue of how it appears within cognition. As a result, knowing finds that what it refers to is not the factor in itself that it took to be its standard of truth. Rather, the object of reference is that criterion *in* its relation for knowing, as it stands defined in the referring underway. By virtue of nothing but its own constitutive truth testing, knowing thus ends up before a new referent, consisting in the being-for-knowing of the former standard of its knowledge. Of course, when what is taken to be in itself gets transformed, the corresponding knowledge cannot remain the same. Since knowledge is knowing's relation to its referent, once the referent changes, so does the knowledge.

In this dual manner, then, the stipulated structure of knowing undergoes what Hegel terms an inversion of consciousness,[10] independently generating a whole new shape for itself with entirely revised poles of truth and knowledge. The referent of knowing has here changed from being what was putatively in itself to become the givenness of this truth *as* it fell within knowing.[11] On the other hand, the corresponding knowledge has changed from being knowing's relation to what was formerly in itself to become knowing's relation to the transformed object.

Nevertheless, to the degree that this emergence of a new shape of knowing has not eliminated consciousness' basic bi-polar structure of referent and reference, the process of knowing's truth testing does not halt. In so far as the new correlative contents *are* differentiated and compared together as the constitutive truth and knowledge of a new shape of knowing, the same inversion process automatically proceeds anew. Since, as Hegel observes,[12] the knowledge of the new object stands contrasted to its object in order to be knowledge for which truth is claimed in the criterialogical, foundational, representational mode of consciousness, the referent once again falls within knowing as something for its consideration. Accordingly, the referent is no longer what is just in itself, but rather what appears to be in itself within and for knowing.

As is evident, the process of inversion will go on unabated so long as knowing persists in claiming truth for its knowledge in the manner of distinguishing what it knows from its relation to it. It matters not whether the referent be construed as sense data, the noumenal essence of sensible appearance, an element of an encompassing conceptual scheme in coherence with which its meaning is determined, a text awaiting interpretation within a hermeneutic circle, or a stimulus causally affecting the behavior of a naturalized cognition. So long as referent and reference are differentiated, with reference relating to an independent factor of any sort, knowing exhibits the structure of consciousness, where the putative standard of truth remains caught in a double bind. On the one hand, the truth criterion, the "in itself", can verify knowledge only by figuring as a transcendent given, enjoying some independence from knowing's relation to it, whereas on the other hand, this same standard can only be appealed to in terms of how it appears immanently within the horizon of knowing. As a result, the standard of truth cannot help but be transformed into a knowledge claim in need of its own truth criterion, setting in motion anew the same tension and the same self-mutation.

It is this unstable, self-transforming relation of transcendence and immanence defining the stipulated structure of knowing that affords phenomenology a method unique among the positive sciences. Due to its process of inversion, the given subject matter of phenomenology stands apart from other objects of positive science not just by making truth claims and verifying its knowledge, but by further determining itself, continually generating new shapes of knowing through the workings of its basic structure. Because the self-examination of knowing produces the development of its own different shapes, whose succession provides what phenomenology considers, the phenomenologist introduces neither any truth criteria or procedures to verify the knowledge of knowing, nor any content at all other than the basic structure stipulated at the start. Non-transcendental phenomenology thus has no need to engage in the arbitrary assertions of eidetic variation, with its dogmatic appeals to the self-evidence of inner intuition to generate new content for some transcendentally privileged structure of intentionality. Instead, the phenomenologist here has nothing left to do but to observe the subject matter as it is given and allow it to develop by itself without any outside interference. Exercising this passive observation, phenomenology attains a non-arbitrary, scientific character of its own so far as all it considers emerges through the inversions of knowing that necessarily follow from the structure stipulated at the outset.

Accordingly, phenomenology must begin with a shape of knowing containing nothing more than the structure of foundational knowing itself. The starting point therefore consists in a knowing where what is known to be true has no other content than that it *is* in itself, that it *is* a given factor to which knowing refers, while the corresponding knowledge has no other content than that it *is* a relation to what *is* given. This entails a shape of knowing whose truth criterion is being and being alone, and whose knowledge is but an immediate certainty of what is. Hegel begins the *Phenomenology of Spirit* with just such a knowing, calling it the shape of consciousness of sense-certainty.[13]

As for what follows, this is already mandated by the phenomenological method: all further content must derive from inversions of knowing succeeding from this rudimentary shape, whose certainty of being will entail an inversion of its own. Of course, this general guideline by no means indicates the actual content of the ensuing succession of shapes. Indeed, whether or not Hegel has accurately described them is a question yet unanswered by Hegel scholarship, which by and large ignores the non-transcendental character of the discourse and treats the description of each shape as if it constituted a true doctrine about the knowing and corresponding objects under view.

Nevertheless, how the succession must come to an end, if at all, can be foreseen from the process of inversion that provides the motor of development. Given how the inversion process is determined through the stipulated structure of knowing, there is only one way that the generation of new shapes of knowing can cease. This is if the referent of knowledge becomes identical to knowing's relation to it. If that happens, then the truth of knowledge no longer becomes something else by getting referred to as something for knowing during knowing's comparison of its truth and knowledge. Since the referent has here acquired the same structure as knowing's relation to it, to grasp it in its relation to knowing is to consider it as it is in itself. Consequently, the content of truth has become completely indistinguishable from that of knowledge, leaving no further comparison to be made.

If such a shape does arise through the successive inversions of consciousness proceeding from sense-certainty, then and only then does the whole process of inversions grind to a halt. Because this process is unique and non-arbitrary, with a definite starting point and a continuous unitary development, Hegel can rightly suggest that the emergence of a shape of knowing where truth and knowledge coincide would signal the completed development of the totality of shapes of knowing.[14] Accordingly, phenomenology would here face its final object, exhausting its own investigation by having observed in the preceding movement every possible manner of making immediate truth claims by referring to what is in itself.

What this leaves is not at all some subject-object identity with absolute knowledge of things as they are in themselves, as Hegel interpreters since Marx and Kierkegaard have commonly maintained. Instead of entailing any such return to metaphysics, the one possible terminus of phenomenology offers a radically novel result, permitting a complete break with the dilemmas of metaphysical and transcendental philosophy that continue to leave contemporary thought at an impasse.

The End of Phenomenology and the Starting Point of a Systematic Philosophy without Foundations

The nature of this breakthrough has been obscured by the fact that Hegel does call the concluding shape of phenomenology "Absolute Knowing",[15] which has unfortunately led many readers to interpret it as the privileged form of philosophical knowing that conceives what truly is. However, if, as Hegel explicitly argues in the *Science of Logic*,[16] phenomenology were to end up presenting a doctrine of true knowing within its own distinct positive science, it would fall into the familiar mistake of stipulating the concept of philosophy prior to the doing of philosophy. This mistake is committed the moment true knowing is rooted in any determinate standpoint. Once this move is taken, as it inveterately is by all transcendental thinkers no matter how they characterize the conditions of knowing, truth is made dependent upon an epistemological foundation that can never be legitimated in its own right, given how all valid claims are assumed to emanate from it.

The true significance of "Absolute Knowing" is better understood by considering what character it must have to be the concluding shape of phenomenology. In the first place, if the stipulated structure of knowing does develop from an immediate certainty of being into a shape where the distinction between truth and knowledge falls away, then much more has occurred than the passing of one shape into another. With the rise of a shape where referring and its referent become one and the same, consciousness has arrived at the point where it itself is forced to recognize that the domain of what is given to it is actually no more than its own posit. With the realm of the given thereby rendered indistinguishable from the reflection of knowing, the entire process of truth-claiming constituting the structure of consciousness immediately collapses. For once such "Absolute Knowing" is achieved, where truth and knowledge can no longer be differentiated, there is nothing left for knowing to distinguish from its own subjective referring as something in itself to which any *objective* knowledge could correspond. Consequently, there can not be any *relation* to a referent, let alone any possibility of claiming truth for such a relation. Without something in itself to which knowing can relate and contrast itself in the dual manner constitutive of the foundational, criterialogical, representational knowing that Hegel calls consciousness, knowing can claim no truth for its knowledge.

This does not signify a supplanting of truth with "warranted belief" or any such version of justification where the standard for adjudicating knowledge claims has a conventional, posited character. These "naturalized" or historically defined criteria remain versions of foundational knowing since referent and reference remain distinct. "Absolute Knowing", by contrast, presents a much more radical outcome. With its equalization of truth and knowledge, where reference has nothing distinct to which to refer, there is simply nothing to be known, nor any knowledge to be held. Consequently, once knowing and what it knows become indistinguishable, there arises no absolute knowledge of what is in itself. What results is rather a complete elimination of truth and knowledge themselves, as they are construed in defining the framework of foundational, criterialogical, or representational cognition. As Hegel observes, absolute knowing is really no knowing at all, but the dissolution of the structure of consciousness,[17] a dissolution that occurs wholly through the efforts of the foundational model of knowing to test its claims and undertake its own critique.

By arriving at this shape of absolute knowing, where there is nothing given to refer to, nor any referring to perform, knowing that claims truth for its knowledge by appeal to an independent referent collapses into literally nothing. Nevertheless, as much as all contrast is removed and no determinate truth claims remain, the one possible conclusion of phenomenology immediately comprises a new point of departure free of the constraints of either metaphysical, transcendental, or phenomenological discourse.

To begin with, since the stipulated structure of foundational knowing has eliminated itself, phenomenology has lost the specific subject matter on which its investigation depends. So deprived of its constitutive object, phenomenology's pure observation is accordingly annulled. Since nothing determinate is left, there is no given subject matter with which any new positive science could proceed to take its place.

Furthermore, since no truth claims remain either about the objects of knowledge or about knowing itself, metaphysical and transcendental discourses have no room for themselves either. With no given about which absolute claims might be made nor any determinate structure of knowing to which authority could be conferred, the metaphysical and transcendental options are set aside.

What there is is neither something in itself nor something in relation to some shape of knowing, but the totally undifferentiated, indeterminate unity into which truth and knowledge have collapsed. As Hegel properly recognizes,[18] absolute knowing's elimination of all knowledge of an in itself has resulted in being, that is, simple indeterminacy, freed of all transcendental conditions and claims to immediate truth. Contrary to prevalent interpretation, this is not being *in res*, the absolute, God before creation, a category of reason, or some transcendentally constituted horizon. Rather, as Hegel repeats time and again,[19] it is utterly unqualified, unanalyzable indeterminacy, which is all that remains when all reference to the given and all correspondingly determinate referring are set aside as a defining framework for arriving at truth.

Although such being is a result of the possible self-elimination of the stipulated structure of foundational knowing, as well as of the phenomenology observing it, its genesis in no way conditions or mediates it. Rather, being is indeterminate and immediate precisely by issuing from a self-annulling mediation. By developing itself to Absolute Knowing, phenomenology serves this introductory role as a process that eliminates itself as a presupposition the moment being emerges from it. Because being here arises with no relation to anything else nor any distinction within itself, there is nothing about it which refers back to any preceding ground or derivation. For this reason, being is really no result at all, but a pure beginning taking nothing for granted nor anticipating anything further.[20]

Paradoxical as it may appear, this sheer indeterminacy is what enables being to provide a remedy for the common impasse of contemporary philosophy. If determinations of any sort were to develop from being without any outside interference, they would comprise the content of a discourse presupposing neither method nor subject matter. Taking nothing for granted, this development of determinations would be immune from the dilemmas of traditional metaphysics and transcendental philosophy, as well as from the relativity of positive science.

In the *Science of Logic*, Hegel attempts to inaugurate presuppositionless systematic philosophy precisely by showing how determinations do emerge from being. Although it might appear inexplicable how anything could arise from such complete indeterminacy, the bare outline of Hegel's argument certainly suggests a possibility.

As should by now be evident, being can be considered as it emerges from phenomenology only if nothing else be admitted. If any other factor were at hand, either as an antecedent ground or a coeval contrast term, the indeterminacy of being would be violated, together with the exclusion of all reference to givens and determinate standpoints from which being has resulted. Therefore, since no other resource can be admitted, any further determination must follow from being by itself, independently of any external positing of either method or subject matter.

Conversely, since this being has no internal distinctions or external relations, it cannot be a ground or cause or determiner of anything, nor can anything arise from it that involves relations or difference to something determinate. Thus, if any other content is to develop out of being, it can only arise utterly groundlessly and be just as undifferentiated and unmediated as being itself.

Although this indicates that nothing can arise from being, Hegel recognizes that nothing does indeed arise from being without any ground at all. As he observes, in so far as being is neither something in itself nor a category of reason, but entirely indeterminate, it is *immediately* nothing, just as nothing is *immediately* the same absence of all form and content comprising being.[21] Consequently, the indeterminacy of being, far from precluding further determination, actually immediately gives rise to a contrast that is no contrast at all, one of being that is nothing and nothing that is being, where each is the groundless emergence of the other.

Indeed, this transition from being to nothing immediately cancels itself as a transition since what emerges from being is really no different than being. Nevertheless, as Hegel recognizes, being has given rise to something other than itself. This is the process of becoming within which being and nothing continually and without intermediary resolve themselves into one another.[22]

If this emergence of becoming suggests how being can be a beginning of presuppositionless determination, of determinacy liberated from the assumption of the foundational framework of consciousness, it also indicates the character of the possible conclusion of the ensuing development. Since what develops from being-nothing-becoming does so without any introjection of given content, be it through reference to what is in itself or what is assumed to constitute true or warranted knowing, the succession of determinations must be an immanent development. In other words, the development from being must be determined through nothing but itself and thereby be self-developing. However, because here the self-development begins with nothing determinate, it is not the self-determination of some content. Such was the case in phenomenology's observation of the self-development of the given structure of consciousness. What proceeds from being is rather self-determination per se.

Hegel draws the necessary conclusion: *what* determines itself from being can only be manifest at the end of the development. Only then has the self-determination fully determined its subject, which is, of course, the development itself in its totality. Thus, being does not comprise the abiding substrate of development, acquiring ever new determination for itself. Being instead actually emerges as the beginning of what finally results only at the conclusion of the development, for at that point that of which being is a beginning first comes into view.[23]

Consequently, what the presuppositionless development from being is a development *of* is left open till the end. Nevertheless, its character can be anticipated in virtue of what is required to bring the self-determination to a close. Since the advance is immanent in character, if the development from being is to come to any conclusion, this can not be certified by any external criterion of completeness or any outside reflection upon the preceding succession of determinations. Instead, the development must itself arrive at a determination that is so structured as to present the interconnection of all the preceding determinations and do so in such a way that they are related together as component elements of a self-determined totality that is both their result and encompassing unity. Such a determination allows the development to close with itself because it not only incorporates all the emergent content within a completed whole, but does so from within the development itself. Accordingly, this final content not only incorporates everything preceding, but renders the development of which it is a part the very process of that incorporation.

However, precisely by being this retrospective ordering of all that has preceded, an ordering in which every content stands as a stage in the concluded self-development containing them all, the last determination is the totality of determinations itself. Hegel calls this final determination the Absolute Idea and appropriately describes how it comprises the resultant self-ordering whole by incorporating all the preceding categories as constituents of its own self-determination.[24] As such, it is the actual subject of the development following from being, comprising what each and every category is a determination of. Consequently, Hegel can rightly say that being is implicitly the Absolute Idea.[25] Furthermore, since the totality of the Absolute Idea provides the ordering principle of its own developed content, it also comprises the method by which all the categories are determined. This is why Hegel can call the Absolute Idea the method of presuppositionless determination.[26]

It can be the method since truth and justification no longer fall apart as they do when the validity of knowledge claims depends upon conformity with an independently given standard, as in the foundational knowing of consciousness. Wherever truth and justification remain distinct, the justificatory process is rendered something outside truth and thereby invalid and incapable of providing any legitimating sanction. This is the basic pitfall of any foundationalism. For if some factor or procedure provides justification for what is to count as true, that justificatory principle cannot enjoy the truth it confers upon knowledge claims since it is given prior to and separately from what it validates. To escape this discrepancy between what it is to possess truth and what it is to confer truth, the justificatory process would have to be determined in accord with itself, which is to say that it would have to be self-determined. In that case, however, its truth would be united with its justification and the distinction of privileged foundation and legitimated knowledge claims would be overcome in the same way in which the attainment of "absolute knowing" eliminates the opposition of consciousness residing in the differentiation of the moments of "in itselfness" and "for itselfness".[27] The positive fulfillment of such a unification of truth and justification is exhibited in the Absolute Idea, whose determinacy owes its truth to itself since what it is is determined by nothing other than itself. The self-grounding process by which presuppositionless determinacy unfolds is thereby nothing other than the self-legitimating account by which truth rests upon itself in express departure from the incoherent dogmatic appeals to extraneously given criteria, conceptual schemes, cultural contexts and the like plaguing the foundation-ridden efforts of so much thought past and present.

If Hegel's strategy be taken seriously and there be granted a presuppositionless development from being, then both its method and subject matter will emerge at the end of the development, instead of being presupposed at the start in the ill-fated manner of positive science, metaphysics, and transcendental philosophy. Whether such a discourse can break through the impasse of contemporary analytic and continental thought must remain an open question until a properly completed phenomenology secures a starting point of being from which follows a completed development of determinacy giving non-metaphysical, non-transcendental philosophy its mandate.

Certainly the predicament of present day philosophy testifies to how it has yet to be shown whether the tasks of phenomenology and of the systematic philosophy without foundations it might introduce have actually been fulfilled by Hegel or any one else. Nevertheless, this same predicament poses the challenge of making these tasks the central problems of philosophy today, while leaving the well-worn path of transcendental argument a thing of the past.

*****

A rare attempt to employ a non-metaphysical, non-transcendental interpretation of Hegel to refute contemporary anti-foundationalism is provided by William Maker's Philosophy Without Foundations: Rethinking Hegel. *A critical review of Maker's argument serves to clarify how a phenomenology of foundationalist thinking can provide access to the presuppositionless startingpoint from which autonomous reason can develop the categories for conceiving reality without foundational assumptions.*

Notes

1 This paper is a revised version of an essay delivered under the title, "Hegel's Answer to the Impasse of Contemporary Philosophy", in October, 1980 at the 6th Biennial Meeting of the Hegel Society of America, held at Trent University, Peterborough, Canada.
2 It is the young Schelling who better fits this description.
3 G.W.F. Hegel, *Logic* (Part I of *The Encyclopedia of the Philosophical Sciences* {1830}), trans. by William Wallace (Oxford: Oxford University Press, 1982), paragraph 78, pp. 111–112.
4 G.W.F. Hegel, *The Phenomenology of Mind*, trans. J.B. Baillie (New York: Harper and Row, 1967), p. 141.
5 Ibid., p. 139.
6 Significantly, Hegel's systematic examination of mind in the *Philosophy of Spirit* conceives consciousness as presupposing and structurally incorporating an unconscious dimension of mind, involving the embodied psyche that relates to its own naturally given determinations without distinguishing them as determinations of an object standing over and against itself. Moreover, the *Philosophy of Spirit* proceeds to conceive consciousness as itself a component of a further dimension of mind that overcomes the defining opposition of consciousness by relating to its mental determinations as both its own and as inherently objective. This provides the psychological possibility for thinking without making the foundational reference to antecedent givens defining the standpoint of consciousness.

7 Ibid., p. 139.
8 Ibid., p. 141.
9 Ibid., p. 141.
10 Ibid., p. 142.
11 Ibid., p. 143.
12 Ibid., p. 143.
13 Ibid., p. 149 ff.
14 Ibid., pp. 135, 137, 145.
15 Ibid., p. 789 ff.
16 G.W.F. Hegel, *Science of Logic*, trans. by A.V. Miller (New York: Humanities Press, 1969), pp. 75–77.
17 Ibid., p. 69.
18 Ibid., p. 69.
19 Ibid., p. 82 ff.
20 For a more extended discussion of how phenomenology can introduce the starting point of foundation-free philosophy without strapping it with a mediation, see "The Route To Foundation-Free Systematic Philosophy", in Richard Dien Winfield, *Overcoming Foundations: Studies in Systematic Philosophy* (New York: Columbia University Press, 1989), pp. 26–31.
21 Ibid., p. 82.
22 Ibid., pp. 82–83.
23 Ibid., p. 71 ff.
24 Ibid., p. 825 ff.
25 Ibid., pp. 829, 838.
26 Ibid., p. 838 ff.
27 For further discussions of the dilemma afflicting any distinction of truth and justification, see Robert Bruce Berman, *Categorial Justification: Normative Argumentation in Hegel's Practical Philosophy* (Ph.D. Dissertation, New School For Social Research, 1983), Chapters 1 and 2; and Richard Dien Winfield, *Reason and Justice* (Albany: State University of New York Press, 1988), Chapters 6 and 7.

Three

Philosophy without Foundations?

Maker's *Philosophy without Foundations: Rethinking Hegel* (Albany: State University of New York, 1994) is that rare work deserving to shift the entire direction of philosophical debate by posing a bold alternative that contemporary discussion has all but ignored. In eleven tightly argued, interconnected essays, *Philosophy without Foundations* presents a radical challenge to prevailing thought first, by defending the option of a non-foundational philosophy that does not abandon objective truth, secondly, by arguing how such a non-foundational philosophy can provide modernity with its only adequate justification, and thirdly, by showing how Hegel's system can be most coherently understood as providing just such a philosophy without foundations. Maker makes these arguments in critical dialogue with an imposing range of thinkers covering modern epistemology from Descartes through Fichte and culminating in the contrast between Hegel's overcoming of foundationalism and contemporary trends from Rorty and Davidson to Gadamer and Derrida. *If* Maker's arguments succeed, he has not only refuted the predominating directions in Hegel interpretation, but, more importantly, rehabilitated the project of systematic, yet foundation-free philosophy, breaking the back of the emerging consensus of analytical and continental theory, and undermining the philosophical basis of the traditional and post-modern opposition to modernity from which a resurgent fascism draws ideological support.

Maker's defense of philosophy without foundations involves a radical reassessment of Hegel's system insofar as Maker shows how the deficiencies of both sides of the contemporary foundationalism/anti-foundationalism debate are exposed and remedied by the defining projects of the *Phenomenology of Spirit, the Science of Logic*, and *Realphilosophie*. Hence, at one and the same time, *Philosophy without Foundations* decisively contributes to Hegel scholarship and to reestablishing Hegel as a living option for philosophy today.

Maker's point of departure is a succinct diagnosis of the terms of this current debate that are so prevalently assumed to be exhaustive, be it by Rorty, Davidson, or Derrida. The modern foundationalism denounced by contemporary anti-foundationalists pursues the Cartesian project of a foundational epistemology employing self-reflection to gain privileged access to conditions of cognition by which knowing can establish the truth of its certainty. [3][1] Anti-foundationalism repudiates this enterprise on the basis that knowing has an opacity resisting any self-grounding self-knowledge due to the ineliminable finitude of cognition's conditioned, perspectival, contextual standpoint. [4] As Maker observes, anti-foundationalists here ascribe a two-fold embeddedness to knowing: 1) a captivity to the *particular* givens of culturally defined practices and 2) a captivity to the *general* framework of givenness endemic to finite subjectivity. Although this characterization allows knowers to free themselves from the particular conventions of a style of reasoning, it leaves them condemned to some foundations of knowing that always remain opaque and unjustified. [5] Hence, when anti-foundationalists expose the "myth of the given", they are merely repudiating the foundationalist privileging of certain *particular* givens [6]; the anti-foundationalists still affirm as an unassailable truth that knowing always founds its claims upon some givens that cannot be rendered fully transparent. [7]

By so absolutizing the finitude of knowing anti-foundationalism condemns itself to a self-referential inconsistency that it cannot defuse. Anti-foundationalism must speak about the unspeakably opaque contextuality of knowing in order to expose the inescapable limits of cognition. Like the foundationalists they denounce, anti-foundationalists unreflectively presuppose that the conditions of our merely human cognition delimit reason, failing to recognize that, as Jacobi noted, the reasoning establishing ultimate limiting conditions of knowing must transcend them, thereby undermining their alleged ultimacy. [34] In asserting that reason is inexorably other-determined, anti-foundationalists join their foundationalist opponents in assuming that reason is exclusively defined by the model of consciousness, that mode of cognition in which whatever is thought always has a given determinacy, fixed in virtue of being a predetermined object of awareness. [35] Although the anti-foundationalists may be justified in denying such heteronomous reason the ability to establish correspondence for its knowledge claims, this only entails the rejection of the very concept of objective truth if truth can only be construed in terms of consciousness. [134] Yet, any such absolutization of the standpoint of consciousness amounts to a surrender to the authority of the given, depriving reason of any critical force as well as the capability of justifying its own impotence. [174] The practical consequences are equally stark: by denying any basis for distinguishing legitimate and illegitimate authority in theory or practice, anti-foundationalism cannot justify its own defense of pluralism and toleration, nor retain the connection between reason and emancipation to which it still often offers rhetorical allegiance. [12]

Maker maintains that the dilemmas of anti-foundationalists reflect the limits of an *ad hoc* deconstruction of foundational epistemology [50] that exercises a *total* negation of the conception of truth as knowable. [191] By arguing that all human cognition is finite and that a self-grounding knowing is impossible *in principle*, *ad hoc* deconstructionists fall prey to self-referential inconsistency that allows the foundationalist to repudiate anti-foundationalism for making unconditional claims about truth and the limits of cognition without being able to justify their own metatheorizing. [50, 53–4]

By contrast, foundationalism can be successfully countered by a systematic, wholly immanent deconstruction of foundational epistemology that achieves a *determinate* negation by observing how foundational cognition's inability to justify its own knowing suspends the absolutization of the scheme of representation, according to which the object of knowing is always already determinate. [50, 191] Thereby systematic deconstruction demonstrates the complementary half-truths of antifoundationalism and foundationalism: whereas antifoundationalism is right in exposing the finitude of foundational cognition, it is wrong in assuming that cognition cannot take any other form, while foundationalism is right in seeking a presuppositionless, self-grounding standpoint, but wrong in seeing such a standpoint as providing foundations for knowing. [58]

To concretize the abiding alternative, Maker turns to the *Phenomenology of Spirit* as the great pioneering undertaking of the systematic, immanent deconstruction of foundationalism from which philosophy without foundations becomes possible. In so doing, Maker provides a solution to one of the perennial puzzles of Hegel interpretation: how the allegedly presuppositionless discourse of *Logic* can still require the prior propadeutic service of the *Phenomenology*. The *Phenomenology*'s immanent critique of foundational epistemology is an indispensable introduction to philosophy without foundations for two reasons. First, the systematic deconstruction of foundationalism is the presupposition for presuppositionless philosophy because it indicates how the absolute beginning of philosophy must have a radically indeterminate and presuppositionless character, free from any external grounding. [77] Secondly, unless the presupposition that all cognition is heteronomous, i.e. predetermined, conditioned by some givens, is removed, philosophy's claim to autonomy would be a "bare assurance", as arbitrary as its denial. [167] In other words, philosophy without foundations cannot proceed unless the epistemological preconception, common to foundationalists and anti-foundationalists alike, that cognition is always conditioned by presuppositions is removed, and removed not by appeal to some counter assumption, but by conditioned knowing's own efforts to justify itself. Consequently, systematic philosophy can be *mediated*, yet make an absolute, unconditioned beginning because what introduces it is the self-elimination of the structure of presupposing [92], e.g. the self-annulling mediation consisting in foundational cognition (a.k.a. the standpoint of consciousness) undermining its own hegemonic claim that reason must always be conditioned. [167] Maker's portrait of this scenario thereby removes the seeming contradiction in Hegel's description of the *Phenomenology* as the introduction to presuppositionless philosophical science.

The question still remains, however, as to what exactly provides the proof of the pudding that the standpoint consciousness cannot justify the exclusive finality of its own finitude. Maker rightly emphasizes that the dilemmas of ad hoc deconstruction can only be avoided if foundational knowing exposes its own limitations. Yet, to present this self-suspension of heteronomous reason, must systematic deconstruction run through all the moves that the *Phenomenology* traverses in journeying from the immediate structure of foundational knowing, sense-certainty, to the shape of absolute knowing, where foundational cognition considers how its cognition of itself can be identical to true knowing?

Maker restricts himself to sketching the final moment, a shortcut that can suffice *provided* that the mere idea of foundational knowing offers sufficient basis for comprehending how its attempt at self-justification must ultimately founder. Maker presents the following argument, which encapsulates the basic moves of the epistemological self-scrutiny that Hegel presents as the shape of "Absolute Knowing": In order to achieve correspondence with the given determinacy that foundational knowing always confronts, foundationalism must circumvent its own defining assumption – that cognition stands in a determinate relation to a given object. [61] For to certify that knowledge and object correspond, cognition is compelled to identify these two terms whose difference enables the comparison to be made in the first place. [62] Only if awareness and its object become equated can the representation of the given correspond to the given. [35] Yet by losing its difference from its object, knowing no longer stands in a determinate relation to an independent given, as cognition must do on foundationalism's own reckoning in order to be more than a subjective projection. Consequently, foundationalism must forego objectivity when its knowledge achieves truth, just as it must forego certainty when it retains objectivity. [63] For when, to adopt the *Phenomenology*'s terminology, consciousness achieves correspondence by knowing its object as itself and itself as its object, that is, by engaging in a pure knowing of knowing, [106] the opposition of consciousness collapses, eliminating its own defining determinate difference between subject and object. Significantly, Maker can make these points without having to concern himself with detailing *how* cognition arrives at equating knowing and its object. No matter what route leads to the attainment of this identity, Maker's argument shows how foundational knowing founders when it thereby meets its own standard for true cognition. The results of systematic deconstruction thus seem to be available without having to follow out the full development of the *Phenomenology*. Maker himself acknowledges that the outcome of the *Phenomenology*'s systematic deconstruction can be understood, but not demonstrated, by considering what consciousness must do to prove that its own structure is definitive of cognition. [89] Strictly speaking, this is achieved by examining not simply how consciousness achieves correspondence, but, more specifically, how consciousness can succeed in knowing its own structure to be determinative of knowable objectivity. Success on this score requires that the *knowing of knowing* be identical to the object of knowing. The shape of consciousness, "Absolute Knowing", with which Hegel concludes the *Phenomenology*, presents just this denouement where the knowing of knowing takes on the character of what is in itself for consciousness and the object in itself becomes determined as the knowing of knowing. The result fits Maker's diagnosis: knowing and object cease to be distinguishable and consciousness forfeits its defining opposition.

If this outcome makes intelligible the critical success of systematic deconstruction, it leaves still to be developed the positive alternative by which the immanent critique of foundationalism comprises a determinate negation. Maker meets this challenge on two principal fronts: first, by identifying the nature of the systematic logic with which philosophy without foundations begins, and secondly, by indicating the limits of the philosophy of reality to which autonomous reason can proceed. In both cases, Maker's analyses focus upon the initial moves of each of these respective branches of systematic philosophy, raising the question of whether such fragmentary treatment can bridge the gap between *Logic* and *Realphilosophie*, let alone sustain the legitimation of modernity with which Maker culminates his argument.

The frugal scope of Maker's positive treatment of philosophy without foundations is determined in the first instance by his strict adherence to considering solely what emerges as the determinate negation of the systematic deconstruction of the foundationalism. What is the positive outcome of the forfeiture of hegemony by heteronomous reason? Maker's answer is, to begin with, understandably programmatic. If philosophy were to begin simply by refraining from presupposing any determinate relation between knowing and its object, whatever would ensue would be a discourse in which both would come to be determined. [64] Any such discourse would achieve self-grounding in the minimal sense of importing nothing determinate from outside itself. [64] Moreover, such a discourse would have no method, if by "method" one means a determinate procedure for considering a given subject matter (i.e. method as construed by foundationalism). [99] Precisely by lacking such predetermination, the cognitive act of autonomous reason would not contribute any determinacy to distort or preconceive what is known. [192] Moreover, since the distinction between knowing and object is not at hand, the insoluble problem of achieving correspondence between cognition and a determinate subject matter cannot arise. [192] Accordingly, autonomous reason does not comprise a thinking about the real, reinstating metaphysical ontology, nor develop the categories by which the real must be thought, in the manner of a transcendental logic. [114] In either case, systematic philosophy would revert to consciousness' structure of cognition, for which even the indeterminate is held as fixed in its indeterminacy as an object distinct from the awareness that thinks it, as already contrastable with the determinate. [110] By contrast, the Logic of autonomous reason develops indeterminacy as determined only in and through itself. [111]

Maker offers only a brief sketch of how this actually occurs, tracing the move from being to nothing to becoming. His account, however, is couched in terms that seem to make appeal to a knowing that strikes different attitudes to the contents it cognizes. Maker begins by construing *being* as what lies at hand when indeterminateness *is thought* as determinate in its indeterminacy. [111] Then, he claims, because the *subjective act* of focusing on *being* as indeterminate invests it with no determinacy, being is equivalent to utterly indeterminate indeterminateness, i.e. *nothing*. [111] Maker then maintains that *thinking* indeterminateness as indeterminate is the *thinking* of a contrast between the minimally determinate and the indeterminate, which, given the absence of anything predetermined, is the *thinking* of a vanishing contrast. [112] What makes the *thinking of* being-nothing-becoming appear to be something standing in a determinate relation to their determinacy is Maker's initial characterization of being as indeterminateness thought as *determinate in its indeterminateness*, a characterization that refers to the *determinate* at a point where such reference can be questioned. The questionability of these details need not impugn the general project of systematic logic, nor the account of *Realphilosophie* to which Maker proceeds.

This account is largely prefigured by the determinate negation of foundational epistemology, since a foundation-free philosophy of reality cannot retreat to any conception of the real that reinstates the opposition of consciousness. The problem of conceiving reality without foundations is framed by precisely what *Logic* does not do. To the degree that *Logic* comprises a self-consideration of thought that assumes neither the difference or identity of thought and reality, it cannot be employed as an absolute method for describing every subject matter. [116, 120] Instead, *Logic* can pave the way for a systematic conception of reality by drawing the limits of its own sphere of pure thought. [116] To think reality without appeal to assurances about "the given", autonomous reason must demarcate what is other to itself by first constituting its limit as pure thought (by achieving logical totality) and then, on that basis, think the idea of a sphere of determinacy that is *not* determined by thought alone. In other words, a foundation-free concept of the real can be arrived at by a self-contrasting or self-externalizing of thought. [118] Accordingly, the distinguishing mark of such non-logical, real determinacies will be that they lack the "internality" or immanent interconnection of logical categories and instead possess an *independence* or *externality*, e.g. *space*. [119–20] This notion of the real does not result from an application of logical categories to any "givens", nor from an assumption of a correspondence between thought and reality. Instead, the concept of reality here emerges from a *self-transformation* that pure thought undergoes in completing its self-determination and delineating its own borders. [120]

Although Maker goes no further in actually developing the systematic concept of reality, he does make bold, and potentially misleading claims about the character of *Realphilosophie* itself, claims that bear upon the relation of theory and practice and the possibility of a justification of modernity. To begin with, Maker spells out the negative character *Realphilosophie* must have to avoid reverting to the pitfalls of foundational cognition: the systematic philosophy of reality can be neither a descriptive account of reality as it is given nor a transcendental deduction of the categories with which reality must be known. [38] *Realphilosophie* is therefore radically nonmetaphysical. [65] Since systematic philosophy does not describe a given world in any of the ways that philosophy has traditionally understood this task, [65] Maker can acknowledge that autonomous reason's successful self-grounding comes at the expense of forsaking the traditional belief that philosophy can say anything *a priori* about the given. [193] Moreover, Maker can consistently maintain that all practical judgments that depend upon assessing what is to be done in given historical circumstances go beyond the purvey of systematic philosophy. [43]

Maker, however, suggests that this departure from traditional dogmas signifies that systematic philosophy has nothing to say about what is (understood as what is given), [194] nor anything to say about the specifics of freedom's actualization. [43] Both of these claims would be acceptable if "what is given" signified solely "what is given as construed by foundational knowing" and if the "specifics of freedom's actualization" referred only to historically contingent conditions. Maker, however, characterizes *Realphilosophie* as purely normative, rather than descriptive, [38, 194] suggesting that a non-metaphysical determination of the given is not possible. Moreover, he claims that a systematic philosophy of right can conceive only the minimal nature of self-determination since any more detailed account would presuppose the contextuality of citizens within the framework of particular institutions and traditions, [43] as if that precluded conceiving the universal structure of the different spheres of right that give freedom its actualization. These intimations, however, may ultimately be matters of emphasis, for Maker acknowledges that systematic philosophy does not deny the reality of anything "given" external to reason; it only bars ascribing evidential primacy to any such givens in conceiving the rational structure of reality. [39]

With this caveat in mind, Maker can address the justification of modernity without transgressing the limits of systematic thought. Indeed, modernity can be justified by appeal to nothing other than the self-grounding discourse of autonomous reason. For, as Maker observes, what distinguishes modernity from all other forms of civilization are claims to autonomy that can only be legitimated by showing how the autonomy of thought and action escape the pitfalls of foundationalism in theory and practice. [14] Far from undermining modern notions of freedom and rationality, the breakdown of the foundationalist strategies of natural right theory and critical philosophy only vindicate the centrality of self-determination that renders modernity free of foundations. [15, 41] Nothing other than philosophy without foundations can justify a world of freedom, for only by overcoming the hegemony of heteronomous reason and its bondage to the given can a genuinely critical, revolutionary philosophy arise with the autonomy enabling it to comprehend modernity's foundation-free embrace of self-determination. [177]

Through the power, clarity and panache of its interwoven arguments, *Philosophy Without Foundations* establishes Maker as a key voice in the most important debates of contemporary philosophy. Whether Davidson, Rorty, Derrida and the legions of post-modernism will confront Maker's challenge remains to be seen. In the interim, it is to be hoped that Maker will waste no time in moving beyond the critical and programmatic arguments he so forcefully outlines to provide the developed positive theory that few since Hegel have dared to deliver.

*****

Although the possibility of an autonomous development of reason may be indicated by the phenomenological critique of foundationalism and the emergence of becoming from the indeterminacy of being, the self-determination of thought cannot be grasped unless the character of conceptual determinacy is itself unveiled. To this end, the following investigation explores how the truth of conceptual determination resides in the connections binding the concept with freedom and with the categories of universality, particularity and individuality

Note

1 Numbers in brackets all refer to page numbers in William Maker's *Philosophy without Foundations* (Albany: State University of New York Press, 1994).

Four

Concept, Individuality and Truth

The Concept in Philosophy

Philosophers may disagree with unmatched zeal, but no philosophical controversy can free them from their common dependence upon the concept as a privileged vehicle of truth. Even when embracing the most radical empiricism or the most unyielding scepticism, philosophers cannot help but make conceptual arguments. All their efforts may aim at curing themselves of addiction to the concept, but so long as they reason on behalf of experience or the suspension of judgment, thinking remains the element in which their claims live and die.

Philosophers' perennial use of conceptual argument might seem a double curse, subverting all of philosophy's equally perennial aspirations to seize the truth. To begin with, if philosophical investigation constitutively employs concepts, how can it possibly establish the validity of conceptual determination? If philosophy cannot fail to use concepts to legitimate reason as a midwife of truth, will philosophical investigation not fall into the vicious circularity of relying upon an instrument that can never be certified without taking its reliability for granted? Secondly, if philosophy must always utilize concepts to arrive at truth, must philosophers not presuppose that the object of true knowledge is what is conceptually determinate? Besides being beyond philosophical justification, will this assumption not limit objectivity to a mortuary of changeless universals from which all becoming, all particulars, and indeed, all actual existence are excluded as some inconceivable illusions? Will the only objectivity that philosophy can conceive be a phantom realm of ideas that can never correspond to genuine reality?[1]

These parallel problems would be fatal to philosophy if the concept were what all too many philosophers have taken it to be: an abstraction whose universality has no bearing upon any other aspect of the particulars to which it pertains.[2] This view has crippling ramifications: First, it renders thinking a formal enterprise that must look elsewhere for the content to which it applies. If the concept is a universal that leaves undetermined the distinguishing features of the many it otherwise unites, everything pertaining to individuation lies beyond the grasp of thought. The problem of the differentia becomes an insurmountable obstacle to the attainment of philosophical truth: for if the concept is a formal universal that cannot lay hold of what individuates its particulars, conceptual determination can never correspond to its objects. The concept may find itself in some property that individuals share, but the individual will always fall outside conceptual grasp. This predicament leaves two equally unsatisfactory options: that parody of Plato, mocked by his own *Parmenides*, where universals are held to be a separate realm of their own against which all individuals are but illusory appearance, or the opposite extreme of nominalism, where individuals devoid of universality are taken to be real and universals devoid of individuality are regarded as subjective illusions.

Either move entails a further consequence of the rigid divide between universal and particular: the freezing of concepts into immobile, isolated abstractions that neither give themselves new determinations nor connect themselves with one another or any other objects. So long as the concept is taken to be a universal that has no inherent relation to its individuation into specific instances, nor for that matter, to its own contrast to different particular universals, it does not become something other than the abstraction it immediately comprises. Instead, it relates only to itself and any relation to something else must be externally imposed. Consequently any thinking that connects such concepts to one another or to individuals will have no basis in the concepts themselves. And if all concepts are of this ilk, any such connections will automatically be bereft of conceptual, which is to say, philosophical justification.

Under these received dogmas the content of concepts becomes just as problematic as their interconnection and objectivity. If concepts neither alter themselves nor relate themselves to what is different from them, their own content has an inescapable givenness. Frozen thus in solitary immediacy, concepts can only be encountered in a series of atomistic receptions, devoid of any intrinsic order or standard of completeness. To the degree that each concept has an isolated givenness, no other conceptual term can possibly certify that a putative concept has a valid content. Every candidate for recollected idea, for example, has no more and no less authority as a conduit of learning than any other abstraction that is immediately offered. Instead of providing truth, concepts get reduced to vessels of meaning and thinking gets condemned to the doxography of analysis.

If philosophy is to go beyond blindly collecting given meanings and instead attain new knowledge through concepts, conceptual determination cannot be limited to abstract universals. This requirement has often been recognized. Plato, for example, identified philosophical thinking with a dialectic in which thought moves from one idea to another without leaving the domain of concepts.[3] Nevertheless, Plato never succeeded in providing an account of concepts that avoided appeal to immediately given contents,[4] nor avoided reducing concepts to forms whose embodiments were incidental to the universal in which they participated. Hence, it can be no surprise that Plato could not provide the self-contained conceptual odyssey promised under the banner of dialectic. Kant, for his part, acknowledged the need for overcoming formal universality both by identifying philosophical wisdom with synthetic a priori knowledge and by making necessary conceptual determination a requirement for the objectivity of representations. As Kant recognized, philosophy could obtain new knowledge of what was necessarily and universally the case only insofar as concepts could be set in binding relation to what was not their immediate identity. Yet, by conceiving such relations as a synthesis, Kant treated the related terms as separate, immobile categories whose necessary relation had to depend upon some third connecting factor – in this case, sensible intuitions as independently given in experience.[5] Far from determining anything individual, concepts remained empty without the external addition of intuitions. Similarly, although Kant allowed the necessary conceptual relation of representations to overcome subjective association and provide for reference to objects, he still left the resulting objectivity relative to appearances. Hence, instead of enabling concepts to lay hold independently of something other, Kant left concepts incapable of determining things in themselves, depriving synthetic a priori knowledge of any unconditioned truth.[6]

The Concept in Systematic Logic

Not until Hegel, and perhaps not since Hegel, has any concerted effort been made to conceive how universality is intrinsically connected to individuality, thereby enabling the concept to surmount the limitations that would condemn philosophy to futility. Despite the growing mountain of discussion of Hegel's work, his contribution on this score has largely been ignored. This is partly due to a lack of appreciation of the systematic program of the *Logic* and partly due to a failure to get beyond the Logics of Being and of Essence and attend to the crowning arguments of the Logic of the Concept.

The connection between these two areas of neglect is far from clear. The systematic program of Logic addresses at one and the same time the challenge of fulfilling the intrinsic demands of logical investigation and the challenge of beginning philosophy without subverting the philosophical enterprise. The answers to these two challenges are one and the same, yet how the resulting solution involves something culminating in a logic of the concept poses questions that have rarely been raised or resolved.

Although logic has continued to be taught and studied as a formal discipline, aiming to uncover forms of thought that underlie the thinking of any content, logic is a thinking of thinking, where form and content, subject and object, method and topic are one and the same. If consciousness be a knowing that constitutively relates to an object it takes to be given independently of its relation to it, then logic can properly proceed only if the defining opposition of consciousness is overcome, that is, if knowing and what is known become indistinguishable. Logic cannot succeed unless its cognition achieves identity with its object because logic will not establish what valid thinking is unless its thinking of valid thinking is itself valid thinking. Accordingly, logic provides the very form of truth, for logical thinking must correspond to its subject matter and do so with a thoroughness that is impossible if knowing relates to something independent of cognition. A discipline that addresses a subject matter distinct from its own thinking cannot help but presuppose its method. Since such a discipline investigates what its topic is, which is something different from the procedure by which it is uncovered, the method is not established by the investigation but must be employed by it as something independently furnished. Moreover, in order to have a specific topic to address with its given method, such a science must presuppose some minimal identity for its subject matter. Otherwise, it has nothing determinate to consider. By contrast, logic can presuppose neither method nor subject matter. Because what logic thinks is indistinguishable from its thinking of it, if logic were to begin with any preconception of its method or topic the identity of valid thinking would be taken for granted instead of being established as the outcome of logical investigation. To avoid begging the question, logic must therefore begin without any determinate method or content. Yet, if logic must proceed without any given method or any given content, whatever content it succeeds in developing will have to be as free of external determination as the form it takes. In positive terms, this signifies that the content must order itself, just as its ordering must be self-determined. Unlike formal universals, whose rigidity requires an external hand to relate them to any further contents, the determinacies of logic must transform themselves and thereby establish the succession of terms that end up comprising the autonomous development of logical thought. Since this free development of self-determined content begins without any presupposed method or subject matter, the self-thinking thought of logic cannot help but be a development of self-determination per se, that is, a development of self-determined determinacy. The indeterminacy of the starting point and the self-determination of the ensuing development are inseparable, for if logic began with any determinate beginning, it would have a determinacy it had not determined for itself. Far from predetermining the course of logical thought, the very project of logic sets logic free of any foundations.

All these implications, which Hegel anticipates in the Introduction to the *Science of Logic*, follow equally from the consideration of with what philosophy must begin. If philosophy were to start with any determinate method it would dogmatically presuppose the form of philosophical investigation instead of considering this as a problem that must be resolved within philosophy itself. By the same token, if philosophy were to begin with any predetermined content, it would take its subject matter for granted, relegating all its subsequent conclusions to claims resting upon an arbitrarily assumed foundation. To begin non-dogmatically, philosophy must therefore start with a complete absence of determinacy, with indeterminacy or being, which can only be thought without mediating qualifications by an equally indeterminate, immediate thinking. If anything is to arise from such indeterminacy, it will have to emerge determined by nothing but itself, for nothing is already at hand to give it a content or an order of presentation. Yet since no foundation can be present as the given substrate of self-determination, what follows from indeterminacy must once more be nothing but self-determination without qualification. Accordingly, philosophy will escape the hold of foundations only by beginning with indeterminacy and proceeding with a self-development of self-determination.

It is not hard to see how the bare outline of Hegel's *Logic* could fulfill the program these considerations anticipate. The tripartite division into successive logics of being, of essence and of the concept can readily be seen to comprise the self-constitution of self-determination. The Logic of Being presents the development of determinacy from indeterminacy. It offers an account of determinacy without further qualification. Of course, if that account began from some given determinacy instead of from an indeterminate starting point, it would beg the question by presupposing determinacy, which is precisely what is to be established. By contrast, the Logic of Essence can and must begin with determinacy, namely the determinacy provided without any presuppositions by the Logic of Being. Without that resource of determinacy in general, the Logic of Essence can hardly unfold the determination of determined determinacy, where determinacy is mediated by some factor that relates to it with a prior immediacy. Only if determinacy is available, can determinacy figure as the determiner of something else whose being it posits. This comprises a two-tiered logic of essence and appearance, ground and grounded, cause and effect and all the further relations involving what rests upon some foundation. Essence's foundational determination sets the stage for the Logic of the Concept to the degree that the latter is a logic of freedom. Freedom's self-determined determinacy presupposes both determinacy in general and determined determinacy since self-determination has determinacy and determines its own determinacy. While incorporating what the Logic of Being and the Logic of Essence provide, the Logic of the Concept transforms both by achieving an identity between what is determined and what does the determining. In this fashion self-determined determinacy emerges as the result of categories that thereby get revealed as way stations incorporated into the process by which self-determination constitutes itself.

This itinerary seems to stand in stark contrast to another division of logic in which the concept looms large. Although Hegel may describe the contents of the three subdivisions of logic in terms that fit the schema of determinacy, determined determinacy, and self-determined determinacy, he also divides logic into an Objective Logic and a Subjective Logic, describing the former as a stage in which the concept is given in its immediacy, prior to coming into its own as a self-mediated totality. Although Hegel qualifies this two-fold division by inserting the Logic of Essence as a connecting bridge, the contrast of Subjective and Objective Logics raises the question of why the concept should map onto the logic of self-determination.

Hegel tells us that he has chosen the term concept with care because the logical determinacies that come under its heading fit the usage that the "concept" has come to have as a privileged vehicle of philosophical cognition.[7] Significantly, when Hegel first introduces the concept as a theme, he initially defines it in terms of two parallel considerations: first, freedom, and then the interconnection of universality, particularity and individuality. The link between these correlative characterizations must be comprehended if the role of the concept in logic is to be unveiled.

Freedom, Individuality, and the Requirements of Conceptual Truth

Hegel introduces freedom as the truth of necessity, describing how the governing relationship of the Logic of Essence resolves itself into self-determination.[8] The basic moves are starkly simple, yet seldom heeded, especially by those who extol the hegemony of efficient causality. The necessitation of determination by some external dominating factor is the pervasive motif of the two-tiered relations in the sphere of essence. With essence, being is always mediated by some foundation. Yet since the foundation has its own determining character only by determining what it founds, what is grounded effectively grounds its own foundation, resulting in a reciprocal interaction where cause and effect simultaneously exchange their roles. Since the erstwhile foundation turns out to have the same determining/determined identity as its erstwhile derivative, the external necessity of foundational determination is supplanted by an internal determination, where the indiscernibility of determiner and determined factor transforms their relation into one of *self*-determination. Although differentiation still occurs, it proceeds within the unity of the self. Since differentiation is now a matter of the self giving itself new determinacy, differentiation takes the form of a development of the self and determination becomes self-development.

That the emergent process of self-determination could have anything to do with universality, particularity and individuality depends upon the conceptual breakthrough that Hegel pioneers in conceiving the relation of these three categories. All the dilemmas plaguing formal universality and the philosophies that rely upon it are supplanted by a simple insight: the universal cannot possess its encompassing unity as a one over many unless the plurality of its particulars can be sustained. Particularity, however, only determines something as an undifferentiated embodiment of the universal. Qua instance, particulars stand in identical relation to the universal in which they participate. Identifying a factor as an embodiment of a universal in no way distinguishes it from any other embodiment. Yet, unless instances can be distinguished from one another, their plurality collapses into one and the universal loses its own constitutive character as a one over many. Consequently, the encompassing unity of the universal depends upon the differentiation of particulars, which is what individuality provides for each. Individuality, as the differentiated particular, is therefore ingredient in universality. And, conversely, because the individual is the differentiated particular, it is equally tied up with universality, without which it loses the particularity that allows it to be more than a classless something or appearance.

The Logic of Self-Determination as the Logic of the Concept

These interconnections between universality, particularity and individuality provide the necessary categorial resources for determining self-determination. Because self-determination must give itself new determinacy to be what it is, freedom minimally involves a successive plurality of determinacies. Since, however, these differentiations remain elements of the self that determines itself in and through them, their plurality comprises a development of the self by which it achieves its self-determined identity. Accordingly, these differentiations are each instances of the self that thereby is their one in many, comprising the *universal* subject that constitutes itself in these its own *particulars*. Since the self can only determine itself by giving itself a *plurality* of differentiations, the particular developments of the self must further be *individuated*. Otherwise their plurality will collapse, eliminating the differences that the self requires if it is to give itself new determinacy. Without the categories of universality, particularity and individuality, determinacy can only be characterized through either contrast with a coeval other or derivation from some antecedent foundation. With a one in many that remains self-identical in the differentiation of individuated instances, the limits of being and essence are overcome and subjectivity, self-development, or freedom becomes determinable.

Granted the traditional association of the concept with universality and with knowledge of what is universal, it now becomes plausible for the logic of the concept to be properly the logic of self-determined determinacy. Because universality is indeterminable without particularity *and* individuality, and because universality, particularity and individuality are the elementary constituents of freedom, the concept and, by way of anticipation, conceptual knowledge are as autonomous as they are universal. Moreover, because universality is inseparable from individuality, the concept need not be skewered upon the problem of the differentia, where formal universality remains incongruent with what it determines. To be trapped within a logic of essence, where determiner and determined never achieve the reciprocity with which self-determination proceeds, may be the fate of the formal thinking of the understanding, whose fixed and rigid representations need external help to connect them with further content. By contrast, if we are to take seriously the concept's baptism in the logic of self-determination, the concept can develop itself, establishing new connections unconditioned by any other factor, laying claim to the title of the synthetic a priori itself.

The Concept, Objectivity and the Logic of Truth

Establishing how the concept is self-determining and self-individuating may set the stage for overcoming the problem of the differentia, but doing so only partially secures the truth of conceptual determination. It enables the concept to acquire a content that is not arbitrary and externally dependent, but it still does not establish how the concept can provide a true determination of something other than itself. So long as the content that the concept gives itself remains purely conceptual in form, it remains subjective in the sense of remaining part of a self whose self-determination does not stand in relation to something else.

In order for the concept to have objectivity, three things must be secured. First, objectivity must itself be established both in distinction from and independence of the concept. This presupposes the development of the concept in its own subjectivity, since objectivity must be other to it. Secondly, objectivity must have a character that is still conceptually determinable. Otherwise, objectivity will remain something forever beyond conceptual grasp. Finally, the congruity between objectivity and the concept must be certified without appeal to some third factor, be it a criterion or a standpoint. If a criterion were employed to assure correspondence, the old sceptical trope of third man argumentation would reintroduce itself, incurring an infinite regress where every application of the privileged criterion requires another criterion to assure its correctness. If a standpoint were required to judge the fit of concept and objectivity an analogous difficulty would be encountered. The above dilemma would, of course, reoccur if the standpoint laid claim to authority by appeal to any criterion. But if the standpoint avoided such an appeal, how could it retain authority against any competing perspective? Since the distinguishing identity of each standpoint would be different from both concept and objectivity in order to comprise a distinct vantage point, nothing in concept or objectivity could unequivocally sanction one standpoint's primacy over another. To avoid this problem, the agreement of concept and objectivity would have to be established from these factors themselves, although the resulting accord would comprise a further irreducible determinacy incorporating them in their correspondence.

These are precisely the central challenges that Hegel's *Logic* addresses in moving from the account of the concept to that of objectivity and that of the idea. These problems are properly handled as logical issues, without appeal to the epistemological frameworks of consciousness or language and the opposition of real selves to the world. If the account of either concept or objectivity were encumbered with any such aspects of nature and mind, the whole enterprise would be sabotaged by the introduction of non-logical determinations whose own difference and conceivability would have to be taken for granted. To escape arbitrariness, the determination of objectivity and its relation to the concept must incorporate nothing but the categories that arise in the self-determination of thought without yet involving objectivity itself.

Objectivity is and cannot be reducible to reality or existence. Reality, as contrasted to negation, is simply the being of determinate being in general. Reality cannot help but be contained in the concept insofar as the concept has determinacy. Admittedly, the concept has determinacy in a manner rather different from how determinacy is contained in the Logic of Being. Whereas the given determinacies of the Logic of Being have determinate being by standing in contrast to other coeval entities, the concept has determinacy that it has imposed upon itself, the determinacy of its self-determination. Unlike given determinacy or determined determinacy, the determinacy of the concept is what it has determined itself to be. Consequently, what is to stand as something other to the subjectivity of the concept must have more than reality. Of course, objectivity will possess reality simply by standing in contrast to subjectivity as something other. But the basis of that contrast must lie in something that is not simply contained within subjectivity; if not, the would-be contrast term gets absorbed without remain.

For just this reason, objectivity cannot simply be existence. Existence has reality, as something determinate, but existence adds the element of determined determinacy common to everything in the logic of essence, where being has lost its immediacy and become founded by some foundation. Existence is determinacy that has a ground and exhibits the relativity of all factors that are determined by something prior to them. The concept incorporates existence to the degree that self-determined determinacy contains determinacy that is determined. Conceptual determination, of course, adds the proviso, issuing from reciprocal interaction, that what gets determined and what does the determining are one and the same. This proviso issues from the reciprocal interaction with which the logic of essence brings itself to an end and introduces conceptual determination. As we have seen, although the two-tiered relations of essence always distinguish between what is founded and what founds, each factor only possesses its defining role thanks to the polar role of its counterpart. Because what founds has its founding character determined by what it founds, a reciprocal determination results where each term is both cause and effect of the other. Since they are indistinguishable, leaving what determines and is determined one and the same, reciprocal interaction itself reverts into self-determination, leaving behind the two-tiered structure of foundational determining and introducing the free self-development constitutive of subjectivity. Although the resulting self-determined determinacy of the concept is irreducible to the determinacy of being and the determined determinacy of essence, the concept incorporates both. What is other to the subjectivity of the concept must therefore be irreducible to each and every category of being and essence.

Instead of being contained within the self of the concept's self-determination, objectivity must be radically independent. Since objectivity can owe its determinacy neither to standing in contrast to other coeval factors nor by issuing from any ground, it must be a self-sufficient totality minimally external to the self-determination of the concept.

Knowledge of objectivity would accordingly escape the limitations of "knowledge" of reality or "knowledge" of existence. Because reality is the determinacy something has in virtue of its contrast to something else, knowledge of reality is always dependent upon knowledge of this other. Yet, if the other only has reality, knowledge of it is dependent upon the same reference to other, introducing the endless dissemination of meaning that Derrideans absolutize as if reality were the ultimate object of knowing.

Knowledge of existence, by comparison, is plagued by the problem of knowing something that is always just the appearance of some determining factor. The latter foundation cannot completely disclose itself in what it determines at pain of collapsing the distinction between ground and grounded by which it is defined. Moreover, if knowing is always only knowledge of what is grounded, no ground can be known as such, since, by definition, the foundation must lack the underlying sufficient reason on which cognition of existence always depends. But then what cognition needs in order to know existence always eludes capture.

To the degree that objectivity is a totality determined in and through itself, objective knowledge cannot reside in contrastive reflections upon what is non-objective nor in attempts to uncover foundations for objectivity. Instead, objectivity can only be known in terms of its own self-constitution, which will give what it is necessarily and with no unencompassed remain. Objectivity is therefore the very entity that could be known without qualification and, in particular, without the quandaries afflicting cognition of reality or existence.

Such a possibility beckons provided knowing can capture objectivity's process of self-constitution. What kind of knowing can lay hold of the internal self-development of objectivity, that is, can grasp what Hegel routinely calls the *Sache selbst*? Neither contrastive or foundational tropes will do. Only a knowing that exhibits the self-determination of the concept can possibly hope to correspond to objectivity's own self-constitution. Precisely because the content of conceptual determination develops itself, genuinely conceptual thought alone can seize objectivity without the distortions of rendering it something relative or foundationally determined by cognition.

Several problems still remain. Objectivity's special suitability for conceptualization might seem to shut philosophy off once more from any grasp of reality or existence. Yet the inability to adequately conceive either reality or existence *independently* need not signify that knowledge of objectivity leaves reality and existence beyond philosophical treatment. To the degree that reality and existence are and revert through their own determinacies into components of objectivity, conceptual determination of objectivity will still consider reality and existence. It will do so, however, in light of how neither reality nor existence can subsist by themselves, nor be known with the same necessity and completeness by which objectivity can be unveiled.

More vexing, however, is the abiding question of just how the concept can achieve correspondence with objectivity. Even if objectivity is determined in and through itself just as the concept is self-determined, objectivity is still other to the concept. If the concept is to be a vehicle for subjectively appropriating objectivity, conceptual determination must have room for both the concept and its other, just as objectivity must retain its difference from the concept and yet be transparent to conceptual determination. Moreover, the accordance of concept and objectivity cannot just fall within *our* observation; if it is to be inherent in concept and objectivity themselves, their determinacies must generate the conceptual structure that contains their correspondence.

Hegel's account of the idea aims to provide just that logic of truth in which concept and objectivity achieve correspondence without appeal to any third factor, be it an immediate reference to reality, as in pre-critical metaphysics, or, recourse to some epistemological structure, as in Kant's transcendental logic, where conditions of sensible experience must be enlisted to determine objects by concepts. Such truth might seem to be of questionable value insofar as the idea contains concept and objectivity as purely logical terms, which lack any resources for distinguishing one or the other from self-thinking thought as something natural or spiritual. Yet this formality is crucial, for if the idea incorporated reference to non-logical terms, their difference and identity would be taken for granted, instead of exhibiting a non-arbitrary agreement of their alleged concept and objectivity. Hence, although the idea incorporates nothing of nature or mind, it supplies the determinacy of truth that anything further qualified as non-logical would have to possess to be known to have an objectivity in accord with conceptual determination.

Significantly, the move to the idea is presented as something resulting simply from the inner workings of objectivity. Instead of depending upon some deus ex machina to set it in agreement with the concept, objectivity progressively transforms itself from mechanism through chemism to teleology whereby it arrives at making itself objectify the form of subjectivity as the outcome of teleological process.[9] Although only a complete determination of the categories involved can certify this pathway, if it does so unfold, it would suggest that the separation of concept and objectivity eliminates itself, that their opposition resolves itself into a unity into which both are absorbed. Any claim that concept and objectivity are incongruent would thereby be shown to rest upon a falsification of their very own identities.

The resultant unity of the idea thereby possesses an origin that would put the lie to such retreats to the opposition of consciousness, for which concept and objectivity are held apart as independently given correlates whose accord depends upon an ever illusive bridge. The idea, by contrast, can be duly identified as the pure form of truth to the degree that it contains concept and objectivity in their own immanent correspondence.[10] What makes this correspondence both objective and conceptually transparent is that, at least on Hegel's self-understanding, each side of the relation has determined itself to be equivalent to its counterpart: through syllogism the concept has given itself determinacy with the self-subsistent totality of objectivity[11] whereas through teleology objectivity has made itself subjective.[12]

The immediate unity of the idea can not, however, be the end of the story. The truth of the idea is first presented as the structure of life, of organic unity. The organism unifies its organs into a conceptual whole, whose universality is the soul animating the particular organs whose own working reproduces their union. What is lacking, however, is a separate conceptualization of the concurrence of soul and body, of the one in many and the totality that exhibits its structure. Without a distinct concept of their correspondence the truth of the idea will not be an object of conceptual determination. If the idea cannot encompass this added feature, then the accord of concept and objectivity within the idea will not itself be in the form of the concept, but will only be conceivable by an independent standpoint of dubious license.

The difficulty is overcome, however, if the idea transforms itself so that the unity of concept and objectivity is contained within it as a conceptual determination corresponding to that unity. Hegel's account of how life develops into theoretical and practical cognition offers a metamorphosis of this sort, arriving at a so-called absolute idea allegedly achieving correspondence between the concept of the idea and the idea's unity of concept and objectivity. The resulting correspondence between the concept of the idea and the idea itself ends up identified as the method of the whole process by which this outcome has arisen.[13]

This notoriously elusive characterization becomes less indecipherable once one considers how the correspondence between the concept of the idea and the idea could not be formulated. Like the correspondence between concept and objectivity, that between concept and idea cannot reside in some third term. To escape the plague of third man argument and other foundational appeals, each side of the accord must posit its counterpart through its own determinacy. If method were an external form applied to an independently given content, method could hardly comprise the process whereby the accord of terms is determined through themselves. The method would instead operate in the traditional manner as an instrument for connecting terms that lack any intrinsic connections. If method is instead the form of a self-developing content, that form will contain the entire content within itself. For if the content is self-ordering, the order of its self-presentation is wedded to the content. Further, since the content of logic, of presuppositionless philosophy, determines itself and since the concept is self-determined determinacy, the form or method of logic will unfold the content in terms of the concept. Indeed, *only* the self-determined factors of the concept could possibly organize the content, for any terms that are not conceptual, i.e. self-determined, would cancel the content's autonomy by making it fit a heteronomous scheme. Accordingly, when Hegel identifies the conceptualization of the idea as the logical method, he proceeds to briefly recapitulate the opening moves of the logic with the added twist of describing how they can be reconsidered in terms of universality, particularity, and individuality, the minimal components of the concept.[14] This does not involve mapping onto logic some independently given schema; it instead comprises a self-organization of the content by terms that emerge within it as it draws to its own close.

Although the details demand investigation like so much before, these concluding moves add a final testimony to why autonomous determinacy and the concept can be one. As all addicts of the concept must admit, the method of philosophy inveterately revolves around conceptualization. The method of logic, of presuppositionless science, is the ordering of the self-ordering of self-thinking thought, an ordering that is defined by the categories of freedom. The concept can be the privileged vehicle of philosophical method because the autonomy of conceptual determination is the anatomy of truth.

*****

The connections between the concept, self-determination and individuality may secure the logical truth of conceptual determination, but they hardly establish how what lies beyond thought should be conceived. How can reason comprehend non-logical reality without reverting to the tainted foundational strategies of metaphysical appeals to the given or transcendental constructions of objectivity? The immediate answer to this question is to be found in conceiving the most elementary factors of nature: space, time and matter. The following chapter takes up this task, drawing upon Hegel's arguments to determine space, time and matter without foundations.

Notes

1 Nietzsche raises both of these objections to philosophy in his *Twilight of the Idols.*

2 As Robert Berman has pointed out in conversation, this formality does not require the universal to be given independently of the particulars to which it applies. As an abstraction, the formal universal depends upon the particulars from which it is extracted even if it leaves them otherwise undetermined.

3 Plato sketches dialectic in his famous account of the Divided Line in the *Republic*, Book VI, 511c.

4 In particular, Plato roots all ideas in the Good, whose given content is immediately intuited, but never shows how specific ideas arise from this foundation, whose own determinacy remains problematic.

5 Hegel makes this point in the *Science of Logic*. See Hegel, Georg Wilhelm Friedrich, *Gesammelte Werke: Band 12 Wissenschaft Der Logik – Zweiter Band: Die Subjektive Logik* (1816), ed. Friedrich Hogemann & Walter Jaeschke (Hambug: Felix Meiner Verlag, 1981), 22.

6 Ibid., 19.

7 See Hegel, G.W.F., *Werke 8: Enzyklopädie der philosophischen Wissenschaften in Grundrisse (1830) – erster Teil: Die Wissenschaft der Logik*, Zusatz to #160.

8 See Hegel, *Gesammelte Werke: Band 12*, 12, 14–15.

9 Ibid., 132.

10 Ibid., 173–4, 176.

11 Ibid., 31.

12 Ibid., 132.

13 Ibid., 237 ff.

14 Ibid., 239 ff.

Five

Space, Time and Matter: Conceiving Nature without Foundations

The Problem of Conceiving Nature without Foundations

Before the juggernaut of scientific achievement and the positivism it fostered, the philosophy of nature has fallen into such disrepute that any thought of rehabilitation seems hopelessly quixotic. Systematic philosophical inquiry into nature has been banished into the limbo of antiquarian scholarship, supplanted by a philosophy of science for whom nature is either the construct of contingently given conceptual schemes or an entirely empirical object equally beyond the grasp of pure reason.

This virtual extinction of the philosophy of nature is not, however, simply a result of the rise of empirical science as a vaunted paradigm of knowledge. It equally reflects the daunting demands that a philosophy of nature makes upon reason. The full challenge of these demands first becomes evident in the aftermath of Hegel's critique of metaphysics and transcendental argument. No longer can it be denied that if reason is to conceive nature systematically three hurdles must be conquered:

First, nature must be determined without appealing to anything assumed to be given in reality. If instead the philosophy of nature proceeds upon a voyage of discovery, offering assurances of encountering certain categories embodied in what it finds immediately at hand, the objectivity of its "findings" and the accuracy of the "descriptions" applied to them will always be uncertain. Not only does any such strategy assume the conformity of nature and reason, but it reads off nature some putative rational content without any argument to legitimate what in particular it discovers. Only a radical departure from reliance upon the given can possibly save the philosophy of nature from the resulting clutches of doubt.

Secondly, nature must not be conceived by appealing to given frameworks of reference, linguistic practices, conceptual schemes or any other structures of cognition that reputedly determine what nature must be to count as an object of knowledge. Any such transcendental constructions of nature fall prey to charges of idealism, where, on the one hand, the determining cognitive structures always stand in need of critique to overcome their own dogmatic assertion, and on the other hand, any resulting construct of nature remains indistinguishable from a subjective/intersubjective stipulation, relative to some assumed framework of reference.

Thirdly, while direct reference to the given and transcendental constitution need to be avoided, the philosophy of nature must still somehow provide categories of nature that go beyond the ideal self-identity of logic's thinking of thought.

Otherwise, the conception of nature will simply recapitulate logic, without capturing the difference by which nature can be distinguished from categories. Yet how can nature be conceived by thought and given a content distinct from thought without reverting to metaphysical assurances about the given or transcendental constructions of a nature receiving its law from the understanding, language or other privileged conditions of knowing?

Hegel poses the problem with stark acuity at the end of the *Science of Logic* by characterizing the gaining of access to the philosophy of nature as a free development of the totality of logical determinacy (the absolute idea), whereby it arrives at self-externality.[1] This notoriously perplexing formulation presents the emergence of natural determinacy as the outcome of a presuppositionless development of determinacy in general that relies upon no reference to what is given in reality nor to any determining structures of knowing. By characterizing the threshold of nature as the self-externality of logical determinations in their totality, Hegel's formulation moreover seeks to arrive at a minimal determination of nature that, on the one hand, makes use of no resources other than the logical categories that have arisen from indeterminacy without assumption, and yet, on the other hand, employs these categories so that they comprise a determination that remains distinct from every last one of them.[2] The difference of nature, by which it is irreducible to thought, is accordingly a pure difference, a difference defined in terms of nothing but thought itself. Hegel has here indicated that since no other resources can be admissibly introduced, the totality of thought can stand distinguished from something other than itself, something more than mere categories, only by arriving at its own totality related to itself as an other. Although each side of the relation may well add nothing to logical determinacy, their relationship, that of self-externality, involves more than either logical totality or any of its component features. Self-externality of the logical totality has this irreducible character simply because it succeeds in incorporating and qualifying in its entirety the whole from which it emerges, without appeal to any illicit extraneous terms.

Yet if the self-externality of the totality of determinacy provides the minimal threshold of something non-categorial, it no more comprises a defining principle of nature than the indeterminate category of being serves as a first principle guiding the development of determinacy that begins from it. If Hegel's project, the project of a fully self-responsible, non-arbitrary philosophy, is to be fulfilled, then the philosophy of nature must treat its starting point, the self-externality of logical determinacy, as the basic component of all further dimensions of nature, presupposed by every one without itself incorporating anything else of nature. Otherwise, this putative starting point fails to be the minimal constituent of natural determinacy, signifying that other unaccounted for contents have been allowed to play a role in nature from the outset.

This prescription imposes an exigence that extends throughout the toils of the philosophy of nature, and indeed, beyond to all further parts of the system of philosophy. In moving from the initial category of natural determinacy to further structures of nature, the conceptualization of nature must always arrive at new terms that incorporate nothing other than what has already been determined and yet comprise something irreducible to their antecedents. Only by following in this path of determinate negation, where the difference between successive determinacies resides in incorporating former terms so as to be other to them, can reliance upon arbitrary stipulation be avoided. Accordingly, what will then certify that the ensuing development is valid is how the subject matter turns out to be nothing but the totality incorporating the determinations that have successively comprised its self-development. In that case, what nature is cannot be detached from the categories determining its self-constitution.

These radical demands of a systematic conception of nature provide a preliminary guide for understanding and evaluating Hegel's initial efforts to determine nature in the opening arguments of his philosophy of nature. There, he begins with space, determined as the immediate self-externality of determinacy,[3] and proceeds to time, motion and matter. Should space come first, as Hegel maintains, followed by time, motion and matter, and should each have the determinations Hegel ascribes to them?

Space as the Minimal Determinacy of Nature

Space will deserve the honor of being that with which the philosophy of nature should begin provided that 1) its own determinacy involves nothing else that is natural and 2) that everything else in nature will involve spatial determinacy. The first consideration can be addressed directly, whereas the second ultimately requires anticipating the entire constitution of nature.

The difficulty of conceiving space resides precisely in the unparalleled elementarity it must possess as the prospective first and most basic element of nature. Hegel identifies space simply in terms of self-externality, ascribing spatial extension little else but a continuity whose differentiation is entirely formal and ideal.[4] Since no factor of nature can be employed to determine space, if space is to be the minimal feature of nature, Hegel departs from the metaphysical approach, canonically followed by Aristotle in his *Physics*,[5] of defining space in terms of place, where existing bodies and their spatial interrelations are presupposed. As Kant recognized, such an approach begs the question by relying upon spatial determinations (e.g. the extension of individual bodies as well as the spatial relations between them) to account for space.[6] However, whereas Hegel grants that space is a pure form of externality, lacking any real individual spatial entities within itself, he departs from Kant by conceiving this pure formality to involve no reference to forms of intuition or any other presupposed cognitive structure.[7] Although in the subsequent *Philosophy of Mind*, Hegel will allow for space, as well as time, to figure as a pure form of intuition,[8] he recognizes that consciousness involves a self-consciousness of a spatio-temporally situated self,[9] which accordingly presupposes space, time and a body of its own within an independently existing nature as a precondition for its own reflection.[10] Moreover, since determining space in reference either to different bodies and their places or to determinate perspectives in the world would take space for granted, distinctions of relative and absolute space cannot enter in at this juncture. Space as such is simply one and unbounded.

As Hegel consistently indicates, the only interruptions in space's infinite continuity are purely ideal points. Points of space are ideal in that their negation of extension cannot rest upon any real thing in space, since space per se cannot be determined by bodies and their relations without falling into a vicious circularity. This leaves no point distinguishable from any other, for each figures in just the same manner as a way station in the continuity of extension. What bounds each point as its spatial other is another point in exactly the same situation, possessing nothing in itself or in its relations to others distinguishing it from its counterparts.[11] Consequently, the point, as the discrete element in space's continuity, generates a line, where one point after another cannot help but pass into identity with the point contiguous with it. Since, moreover, the resulting line is just as indistinguishable from the edge of space it terminates, space's internal differentiation into point and line entails planes, whose own analogous similitude fills the continuity of space with the unbounded extension of spatial dimensions, which are equally indistinguishable from one another.[12] It makes little difference whether height, length and breadth be construed in Euclidean space or in a non-Euclidean space,[13] whose use of curvilinear axes already presupposes the two and three-dimensionality defined by straight lines.[14] In each case, what a dimension is identified as is a matter of indifference, since every one can substitute for the others without altering its internal or external relations.[15]

Be it situated as a point on a line, a line on a plane, or a plane within spatial dimensions, each "here" accordingly remains utterly identical to every other, comprising, as Hegel puts it, merely the possibility of a determinately identifiable place.[16] Other differentiating factors must be introduced if any here is to be become an actual place, abidingly distinguishing one space from another, and therewith establishing the resources for contrasting relative and absolute space.[17] Yet what are such real spatially individuating factors and how are they to be accounted for without arbitrary assumptions?

Hegel's own solution is perplexing at first glance. If one follows the course of his argument and the order of constitution in which categories are introduced and determined, the spatial individuation of place arises out of a unity of space and time, which itself first requires that time be accounted for following space.

Time as the Truth of Space

The treatment of time as the conceptual successor of space involves two complementary claims: 1) that space is determined independently of time and 2) that time requires space and nothing but space for its own determination.

The first claim is already confirmed by how the determination of space as sheer extension, i.e. immediate self-externality, involves nothing but the continuity of ideal points, whose passage into lines, planes and all-encompassing dimensions proceeds apart from any temporal differentiations.

However, how time can follow from space appears inscrutable, if not preposterous, on several grounds. On the one hand, the very abstractness of space seems to offer hardly anything with which temporality could be constituted. On the other hand, although time may well presuppose no other natural factors in its own process, how the passage of past, present and future requires space is hardly evident. The familiar appeal to motion as a basis of time may employ changes in place, but since such changes comprise motion only in conjunction with temporal alteration, motion must be derivative rather than constitutive of time.

Accordingly, the temptation is great to treat space and time as two independently given factors, each as inexplicably basic as the other, and only externally connected in motion.[18] Kant, for example, initially appears to grant time an equiprimordiality with space not only by describing it without reference to space as a pure form of intuition in its own right, but by identifying it as the form of inner experience, which seems to operate without space at all. Of course, in his "Refutation of Idealism", Kant later acknowledges how internal time consciousness is predicated upon external spatial experience to the degree that temporal duration can be experienced only if consciousness can reidentify something enduring as an objective backdrop for the subjective succession of its own representations.[19]

However, how can time be predicated upon space if space and time are to be determined in themselves without reference to a structure of cognition whose own individuality ultimately depends upon being an embodied standpoint, already ordered in space and time?[20]

Hegel's answer is brutally simple: time follows from space insofar as time is what space becomes.[21] With spatial continuity involving a negation of extension in the point, which no less extends into other points, whose encompassing line both terminates and extends into adjacent lines, forming planes and dimensions, space exhibits an inherent contradiction: wherever spatial limits occur in the constitutive continuity of extension, they become external to themselves, transgressing the boundary that separates them from what they demarcate. They thus cancel their own discrete character on which space's expanse itself depends.[22] For if points, lines and planes cannot hold themselves apart, externality collapses. Yet, as continuous, space must involve both the differentiation of point, line and plane as well as their self-transgression. Hegel claims that time consists in just this ongoing self-transcendence endemic to space.[23]

If this is the case, then space directly generates time as a qualification of itself and time inherently refers to space by being that wherein extension and its negation are at one.

Yet why should the process of space count as time?

To begin with, it might appear that the ceaseless self-transcendence of spatial differentiation is no more than the negations within space whereby points, lines and planes comprise the intra-spatial terminations of extension. However, if space as a whole transcends itself, the negation in question cannot fall within space, as a merely spatial limit. In this sense time can count as the self-transcendence of space in its entirety, for the differentiation between space at one moment and at another is that whereby space as a whole is external to itself. That differentiation goes beyond any distinguishing between points and lines and planes that is internal to space. Instead, it comprises a negation of space in which space stands in relation to its other, where that other is defined in terms of nothing but the totality of space posited as external to itself. Since each differentiated space is subject to the same self-externalization by which each spatial now is immediately supplanted by another, time's negation of space is ongoing. In this way, time becomes determined by relying upon no other resource than space and yet achieves irreducibility to spatial differentiation. Time is, in other words, the self-externality of immediate self-externality, just as space is the self-externality of logical determinacy.

Consequently, time cannot be determined independently of space, for it is nothing but the process whereby space is differentiated from itself. Each halt of this process is the entirety of space, where all spatially connected "heres" get further subsumed under the determination of "now", setting them external to space at the next moment.

At every juncture, the mere self-externality of space provides nothing other than this abstract difference of space transcending itself, where each successive space has nothing in its own spatial character to distinguish it from its counterparts, which themselves immediately pass over into an otherwise indistinguishable spatiality, which undergoes the same self-transcendence. Only when real motion enters in, requiring matter and its differentiation into bodies in space, can spatial configurations at one moment be actually distinguished from those at another. Since, however, motion presupposes the passage of time, time itself cannot involve a self-externality of space relying upon the movement of bodies to distinguish space from itself at different moments. Moreover, matter and its diremption into bodies must itself be constituted, which hardly can occur on the basis of space alone.

Owing to the absence of any such real differentiation as bodies in motion could provide, time's process is just as continuous as the intra-spatial differentiation of extension.[24] Nothing real separates one now from another except the externality of space at this moment to an otherwise indistinguishable space at another moment. The differentiation of one moment from another is thus nothing but the self-externality of space in its entirety. Accordingly, just as space as a whole perennially becomes self-external, so time continually consumes its own moments, immediately supplanting one now with another equally straddling past and future. Whereas space surmounts its internal boundaries by continuing into other points, lines and planes that remain external to one another, time cancels each moment that comprises the other or negation of its predecessor, allowing Hegel to characterize time in terms of a negation of the negation or self-relating negation.[25] As this self-devourer, time never exists as anything more than space, just as space never is without subjecting itself to time, acquiring a transient presence arisen from the past and succumbing to the future.

Lacking any further real differentiation of one temporal moment from another, time is as abstract and ideal as space,[26] equally comprising a mere form of nature, awaiting the concretizations that further natural factors could bring. Once more, this very formality leaves no room for injecting the very concrete distinctions between objectivity and subjective consciousness that Kant will add by treating time as a pure form of subjective intuition.[27] When consciousness takes its place in the world, all its intuitions may well be temporally ordered, but this hardly reduces time itself to nothing but a fixture in the mind.

Nevertheless, the preceding basic determinations of time as the outcome of space appear to be flatly contradicted by the temporality of internal consciousness, wherein representations follow one another in time in a sphere presumably devoid of extension. If indeed we have representations before and after one another without setting space external to itself, time can hardly involve, let alone issue from, space. Yet if the sequence of representations involves no reference to space, how is their succession distinguishable from a purely logical ordering? Kant's appeal to a spatial backdrop in his "Refutation of Idealism" reflects this inability of keeping temporal succession from collapsing into logical succession when space is not incorporated into the determination of time.

Indeed, it is time's constitutive reference to space that permits the sequence of the *categories* of space and time to be, like that of all categories in the philosophy of nature, a merely conceptual ordering. The move from point to line to plane is not temporal in character because it itself involves no continuous differentiation of spatial backdrops. A hopeless paradox would, of course, arise if the transition from space to time was temporal, rather than categorial, for then the very emergence of time would be preceded by a passage of time.[28]

The Determination of Motion and Matter from the Unity of Space and Time

Precisely because time incorporates space as a constitutive factor in the perennial movement of the future into the present and the present into the past, time itself involves a space-time continuum in which each spatial point is no less situated at a moment in time. By so uniting the here and now, time provides the resources for distinguishing "heres" by their different times and for distinguishing different times by their different "heres". Moreover, time's joining of here and now permits a point in space to retain its position over time and thereby be a temporally enduring point, distinguishable from others. Hegel appropriately calls the immediate unity of space and time, "place", insofar as it provides a concrete space-time location sufficient to provide a positive existence for differences of both time and space.[29] Without any connection to temporal moments, each point in space is indistinguishable from its counterparts, just as each moment in time is indistinguishable from its successors without any tie to a particular here. With the joining of place, however, time can have a determinate duration and space can have determinate locations that persist, instead of disseminating themselves into indifferently juxtaposed points.

Moreover, with place providing enduring, reidentifiable points in space, as well as a positive existence for the passage of time, motion and its counterpart, rest, become determinable. With identifiable spatio-temporal locations, the ideal self-transcendence of point into line and line into plane can be tied to the succession of temporal moments, constituting motion as the double-sided becoming where time transpires in terms of a succession of different places and space extends itself in terms of the sequence of time.[30]

Although the particular spatio-temporality of place provides the resources for determining the presence and absence of motion, movement and rest remain wholly ideal since nothing is available to distinguish moving from stationary points except a wholly external and arbitrary reference to other points that incorporates them into some trajectory in time. By themselves, space, time and place provide no basis for connecting one point rather than another to its adjacent counterparts, making one motion more actual than any other possible alternative. Consequently, the possibility of time embodying itself in a succession of different places and of space having identifiable places in time remains completely formal. For this very reason, as

Hegel observes,[31] each place immediately passes over into another place, otherwise indistinguishable from itself, generating formal motion, just as each point passes over into adjacent points to generate a line.

What can overcome the ideality of motion and the formality of place in space-time is, of course, matter, differentiated into a plurality of bodies determinately occupying space in motion and at rest. Yet how is matter and its bodies to be accounted for on the basis of the only legitimate resources at hand: space, time, place and formal motion? The difficulty of this problem might appear to be easily escapable by simply removing the restriction to such meager constituents. However, what other natural factors could possibly be appealed to that do not themselves already involve matter?

The logical account of determinate being faced the analogous difficulty of having to constitute determinacy without employing any factors that are already determinate. Determinacy could accordingly only be comprised of a unity of being and non-being, for any other solution would beg the question by using determinacy, which is precisely what is in question.[32]

To avoid a similar vicious circularity, matter must be constituted without any material factors. Consequently, the very ideality and formality of space, time, place and formal motion qualify them as resources in nature that can serve to account for matter. Yet how can the immateriality of space, time, place and formal motion become constitutive of the basic materiality incorporated in all further dimensions of nature?

Hegel's answer is as simple and confounding as his parallel solution to the emergence of determinate being in logic: just as determinacy arose as the unity of being and non-being issuing from the internal collapse of coming to be and ceasing to be in becoming, so matter is determined as the unity of place and motion that arises from the internal collapse of motion's passing away and regeneration of space in time and time in space.[33] In motion, the passage of time presents itself spatially as the move from one place to another, where time continually reverts to space, while space simultaneously differentiates its extension in a temporal succession, extinguishing itself into time. Just as what becomes is something, as a paralyzed unity of being and non-being, so what moves is matter, as a unity of space and time where a determinate extension persists, brought along by time in its passage through space. Place, the spatial now, becomes united with motion to the extent that the temporal succession of time *both* renders place external to itself (as passing into another place which passes on into another, and so forth), and retains the self-identity of place along this trajectory of ongoing self-transcendence. In other words, a determinate, reidentifiable space travels along the trajectory of motion.

In so doing, two constitutive features of matter are exhibited: the occupancy of a determinate space and the enduring character of that occupancy through change of place in time. These coordinate aspects of matter depend upon both space and time for their constitution. If instead occupancy of a determinate space were instantaneous rather than enduring, materiality would vanish, just as it would if occupancy of a determinate space did not persist over change of place. For this reason, matter must exist in time, as well as in place, and be movable.

Whereas the movability of matter presupposes its spatio-temporality, it might be thought that the existence of matter in time and space need not be tied to movability. After all, why could not all motion be external to matter, as simple mechanics might be thought to presume, not only allowing for inertia, but allowing for matter or, at least, some matter to be immovable? In that case, motion would either be an inexplicable accident or the result of a mover beyond matter. However, what precludes the immovability of matter are not the problems of such ramifications, but rather the inability to distinguish matter from mere space if matter cannot maintain its extension while changing its place. This persistent occupancy of space during motion is a key element in any dynamic conception of matter, which conceives of material extension in terms of moving force and thereby conceives matter and motion as ineluctably connected.

Not surprisingly, Hegel adopts such a dynamic conception of matter, further determining matter as a unity of attractive and repulsive force. In so doing, he follows Kant, with the qualification that Hegel jettisons reference to conditions for the experiencing of objects as well as eliminates the external relation in which Kant leaves the forces of attraction and repulsion in matter.[34] Reference to the framework of experience cannot be tolerated since it takes for granted cognitive structures that themselves presuppose a subject bodily existing in the world. The forces of attraction and repulsion cannot be treated as independent, separately given aspects of matter because the dynamic materiality they constitute cannot involve one without the other. As Hegel points out in his critique of Kant in the account of one and the many in the *Science of Logic*,[35] attractive force operates only insofar as what is attracted has not collapsed into a single point, but remains self-repelling, whereas repulsion operates only insofar as what is repelled has not dispersed to infinity, but retains a self-attraction resisting its dispersal. Consequently, if matter occupies a determinate space through moving forces, those moving forces must involve both attraction and repulsion.

However, if the coeval operation of attraction and repulsion permits matter to occupy space by excluding similarly exclusive material, it threatens to coalesce into an endless homogeneous expanse of matter. This would result in all the conundrums of Descartes' equation of matter and extension,[36] wherein degrees of density become a mystery and the possibility of actual motion is eliminated, since the space that matter "leaves" or "enters" might as well be occupied.

Matter can only uphold its own durable and reidentifiable occupation of space in time if, as Hegel's argument suggests,[37] matter constitutively possesses a center of gravity, allowing matter to be differentiated into a plurality of real bodies, whose attraction and repulsion as discrete masses is mediated by an independent unity of attraction and repulsion holding each body together as something in itself in its relation to others. This minimal play of gravity provides, as Hegel aptly puts it,[38] the substance of each body, maintaining internal cohesion and independence from its counterparts during motion and rest, as well as compression and expansion. Gravity does so, as would be required, by employing nothing but attraction, repulsion and the relations of space and time that they incorporate. For when attraction and repulsion are combined, preventing the collapse of matter into a single point, attraction figures as gravity, drawing matter to a center that matter's own repulsion prevents it from attaining.[39]

Hence, gravity is not external to matter, but the immediate result of the combination of attractive and repulsive force basic to material being. Moreover, since possessing a center of gravity renders matter a real unit, forming a body able to interact as such with others,[40] the differentiation of matter into bodies is equally inherent in material nature. Furthermore, by providing the minimal independence and discreteness of individual bodies, whereby bodies can act upon and resist one another, gravity equally entails the elementary relations of external mechanical motion: pressure and impact.[41]

Avoiding appeals to the given or transcendental constructions, Hegel has here succeeded in being a midwife to the concept of nature, presenting the self-emergence of its primary determinations: space, time, motion and matter. If the philosophy of nature is to regain respectability, it must follow this path of conceiving nature without foundations, leaving behind the fast and loose adventures of its metaphysical and transcendental past. Only then will the Einsteins of tomorrow have something more to hang their theories on than the latest absence of conflicting observation.

*****

Conceiving what is and conceiving what ought to be each require emancipation from foundations and a corresponding enlistment of autonomy. The parallels applying to justification in theory and practice might appear to overshadow all differences, especially when legitimate conduct is conceived as ethical community, where ought and is seem to converge. Confronting the lure of ethical community serves as a suitable bridge to the theory of right because exposing the shortcomings of communitarian thought and redeeming ethical community as an institution of freedom sets in relief the basic structure of legitimate practice.

Notes

1 Hegel, G.W.F., *Science of Logic*, trans. by A.V. Miller (Atlantic Highlands, NJ: Humanities Press, 1989), 843.
2 In so doing, Hegel seeks to obey the two fundamental requirements of any immanent categorial development: 1) that each advance to a new category relies upon nothing other than what has already been established, and 2) that each advance nevertheless achieves its own irreducible identity through that means. Such an advance can only occur if categories entail their own determinate negation.
3 Hegel, G.W.F., *Philosophy of Nature*, edited and translated with an introduction and explanatory notes by M.J. Petry (London: George Allen and Unwin, 1970), paragraph 252, I:217.
4 Hegel, *Philosophy of Nature*, paragraph 254, I:223.
5 Aristotle, *Physics*, Book IV, Chapter 4, 210b35 ff.
6 Kant, Immanuel, *Critique of Pure Reason*, trans. by Norman Kemp Smith (New York: St. Martin's Press, 1965), A32/B38, 68.
7 Hegel, *Philosophy of Nature*, remark to paragraph 254, I:223.
8 G.W.F. Hegel, *Philosophy of Mind*, trans. by William Wallace and A.V. Miller (Oxford: Oxford University Press, 1978), paragraph 418, p. 159. Significantly, Hegel here does not simply find space and time as forms of intuition, as Kant does, but instead argues that intuition gives its content the form of space and time because mind here relates to its own determinacy as external to itself, entailing the same type of self-externality by which the culmination of logical determinacy comprises the minimal threshold of nature.

9 In this way, Hegel foreshadows aspects of Strawson's argument in *Individuals* (New York: Routledge, 1990), with the qualification that Strawson remains committed to a transcendental construction of nature. As a result, Strawson is plagued by the difficulty of upholding the primacy of our conceptual scheme in defining the reality given to us while acknowledging how our awareness is itself predicated upon the spatio-temporality of our being in the world.
10 For this reason, Augustine's account of time as a mental projection begs the question. Indeed, Augustine's own argument makes this clear insofar as he must presuppose the duration and temporal ordering of mental representations in order to give his constructions of past, present and future in terms of memory, attention and expectation. See Saint Augustine, *Confessions* (Harmondsworth: Penguin Books, 1979), Book XI, sections 27–28, pp. 275–278.
11 Hegel, *Philosophy of Nature*, addition to paragraph 254, I:224.
12 Hegel, *Philosophy of Nature*, addition to paragraph 254, I:224.
13 Eva T. Brann gives a telling argument that insofar as only Euclidean geometry permits "the preservation of shape across variations of size", our imagination, for which imaging similarity is essential, must operate in terms of Euclidean space. See Eva T. Brann, *The World of The Imagination* (Savage, Maryland: Rowman and Littlefield, 1991), 604.
14 Hegel points out that curvilinear lines are derivative of straight lines, since curved lines involve two dimensions. See Hegel, *Philosophy of Nature*, addition to paragraph 256, I:227.
15 Hegel, *Philosophy of Nature*, paragraph 255, I:225.
16 Hegel, *Philosophy of Nature*, addition to paragraph 254; I:224.
17 As Hegel points out, relative space involves the determinate spaces of bodies, whereas absolute space is the sheer abstractness of space, as it extends with indifference to every body within it (Hegel, *Philosophy of Nature*, addition to paragraph 254; I 225).
18 Hegel points out that space and time are ordinarily taken to be completely separate, so that space is given by itself and then, in an unrelated move, time is "also" to be found. See Hegel, *Philosophy of Nature*, addition to paragraph 257, I:229.
19 Kant, *Critique of Pure Reason*, B275–B279, 245–247.
20 This is the question that escapes, and cannot help but escape, such transcendental thinkers as Strawson and Heidegger who conceive the conditions of knowing as a subjective being-in-the-world.
21 *Hegel, Philosophy* of Nature, paragraph 257, I:229.
22 To paraphrase Hegel in paragraph 260, space is the contradiction of indifferent juxtaposition and continuity devoid of abiding difference (Hegel, *Philosophy of Nature*, paragraph 260, I:236).
23 Hegel, *Philosophy of Nature*, addition to paragraph 257, I:229.
24 Hegel, *Philosophy of Nature*, remark to paragraph 258, I:229.
25 Hegel, *Philosophy of Nature*, addition to paragraph 257, I:229.
26 Hegel, *Philosophy of Nature*, paragraph 258, I:229.
27 Hegel, *Philosophy of Nature*, remark to paragraph 258, I:229.
28 Similar difficulties trouble Augustine in considering the temporality of creation.
29 Hegel, *Philosophy of Nature*, addition to paragraph 260, I:236.
30 Hegel, *Philosophy of Nature*, paragraph 261, I:237.
31 Hegel, *Philosophy of Nature*, paragraph 261, I:237.
32 For an analysis of this constitution of determinate being, see Chapter 3, "Conceiving Something Without Any Conceptual Scheme", in Richard Dien Winfield, *Overcoming Foundations: Studies in Systematic Philosophy* (New York: Columbia, 1989), 55–75.
33 Hegel, *Philosophy of Nature*, paragraph 261, I:237.
34 For Kant's account, see Immanuel Kant, *Metaphysical Foundations of Natural Science*, trans. by James W. Ellington, in Immanuel Kant, *Philosophy of Material Nature*

(Indianapolis: Hackett, 1985), Second Chapter, "Metaphysical Foundations of Dynamics", Ak. 496–535, 40–94.

35 G.W.F. Hegel, *Science of Logic*, trans. by A.V. Miller (New York: Humanities Press, 1976), 178–184.

36 Descartes, *Principles of Philosophy*, Part II, sections 4, 11, in *The Philosophical Writings of Descartes*, trans. by J. Cottingham, R. Stoothoff, and D. Murdoch (New York: Cambridge University Press, 1985) Volume I: 224, 227–228.

37 Hegel, *Philosophy of Nature*, additions to paragraphs 265 and 266, I:248,251.

38 Hegel, *Philosophy of Nature*, remark to paragraph 262, I:242.

39 Hegel, *Philosophy of Nature*, addition to paragraph 262, I:242.

40 Hegel, *Philosophy of Nature*, addition to paragraph 266, I:251.

41 Hegel, *Philosophy of Nature*, addition to paragraph 265, I:248.

II
Right

Six

Ethical Community without Communitarianism

The Rationale for the Renewed Turn to Ethical Community

In both theory and practice, controversy over framing conduct and institutions has increasingly revolved around the idea of ethical community. Advanced more and more frequently under the banner of "communitarianism", this idea has cast an alluring spell. However vaguely it be drawn, ethical community is widely entertained as a practical alternative to the alienations and anomies of modern life, where instrumental rationality and atomistic self-interest have allegedly supplanted tradition without supplying any new bonds of intrinsic worth. And just as commonly, the concept of ethical community is appealed to as a remedy to the complementary dilemmas of the two leading competitors in contemporary ethical theory, teleological and procedural ethics. On both fronts, a turn to ethical community has taken center stage. If this turn sometimes has a nostalgic air, yearning for the faded spirit of ancient assemblies, town meetings, and village commons, it equally reflects the perennial character of the underlying issues, as much in dispute when Plato and Aristotle argued for the polis in opposition to the natural bonds of tribal association and the instrumental association of social contract, as when Hegel defended the ethical life of family, civil society and state against the advocates of restoration and the contractarians of his day. Although we can take advantage of the experience of past ages that have grappled with ethical community in theory and practice, the worth of their achievements can only be known by first independently certifying the valid content of ethical community and the place it should command in the conception and reality of conduct.

The current rehabilitation of ethical community, comprising the communitarian turn in ethics, has its own lessons for reconstructing theory and practice provided one rethinks its formulations with the same critical autonomy. Above all, what the communitarian turn uncovers of the inherent logic of ethical community must be distinguished from what it ignores and perverts.

To its credit, the communitarian appeal to ethical community is grounded in an understanding of the complementary limitations of teleological and procedural ethics. Admittedly, teleological ethics adopts a plausible strategy by seeking the legitimacy of conduct in the given nature of its ends, transforming ethics into a science of a highest good. Yet every candidate for a highest good is subject to the skeptical doubt that its particular content is devoid of universal validity. Since any putative ultimate end, be it happiness or human perfection, cannot retain its normative primacy if any other factor grounds its archimedean role, its authority must reside in its own immediate givenness. However, every particular content can be ascribed an immediate givenness, especially when no higher tribunal can be relied upon without undermining the primacy of the end it is intended to sanction. Hence, teleological ethics remains caught in the bind of rooting conduct in a particular content, whose universality remains doubtful.

By contrast, procedural ethics falls into a precisely inverse dilemma. Abandoning any attempt to found normativity in privileged ends, procedural ethics instead appeals to a privileged form of willing, involving a principle that determines conduct independently of any particular end. Right hereby precedes the good, insofar as what makes conduct valid is conformity to an antecedently prescribed formal requirement that first allows an end to qualify as good. Yet, because the privileged form of willing is prior to all particular ends and actual conduct requires that something particular be willed, this content must derive from some other source. But no further law can mandate how to decide which extraneous end should be chosen as the goal of action in conformity with the privileged procedure, since the procedure has exclusive normative primacy. As a result, whatever end is willed enters in arbitrarily. For just this reason, procedural ethics is haunted by the specter of formalism, according to which its formal principle of willing cannot mandate any particular end and determine actual conduct unless some privileged good is surreptitiously introduced. As a consequence, the ethical remains an unrealizable ought to be and the agent remains caught in an interminable struggle to act in independence of given ends without benefit of any unequivocally sanctioned alternative. Hence, whereas teleological ethics stamps conduct with a particularity lacking universality, procedural ethics leaves conduct governed by a universal law devoid of any non-arbitrary particular filling.

On both sides, the root of failure has a common logical form: an inability to unite the particular and universal dimensions of conduct. Accordingly, overcoming the limits of each side involves a common solution: somehow arriving at a form of willing whose universality is concrete, possessing a lawfulness that specifies particular ends of its own, or, to put it inversely, involving action in pursuit of particular ends that intrinsically enjoy a common legitimacy.

It is just such a solution that the communitarian turn purports to find in its idea of ethical community. Defined in function of the remedy it is to provide, ethical community is here conceived as an association in recognition and pursuit of shared values, where membership involves fulfilling a role carrying with it the obligation of consciously reproducing the bonds of community that comprise the very association in which that role can alone be performed. On this basis, each member is able to pursue particular aims that are automatically shared in common by all other individuals who duly fulfill the roles defined by their membership in the community. Alternately, what is recognized by all individuals as a condition for belonging to and participating in the community is a common good that contains within itself the activities that realize it. That is, the structure of the community is itself what all participants realize in reaching the aim of their role, just as their activity is nothing but a constituent element of the association they perennially seek to sustain in fulfilling their station.

This intrinsic connection between the agency of membership and the unity of the community overcomes the onesidedness of teleological ethics insofar as the particular ends that are sought in occupying one's station are inherently universal with respect to the community. Indeed, their conformity with the ends of every other member consists in nothing other than the reproduction of the unity of the community, which exists in nothing but their congruent activity.

On the other hand, the limits of procedural ethics are equally overcome in that the universality of conduct consists in acting in conformity to actual roles that presuppose and reproduce the very good that they seek to realize. What ought to be is no longer a beyond, but rather a form of community that must already be at hand in order to make possible engagement in the very roles that aim at its reproduction. Accordingly, each agent is no longer caught in the dilemma of choosing between arbitrary goals and conformity to an empty lawfulness, where morality is an endless striving, always falling short of the fulfillment of duty. Instead, each member of the community actually achieves ethical virtue by fulfilling his or her role, whose fulfillment is the very life in common comprising the community to which each belongs.

The Fatal Limits of the Communitarian Turn

From F.H. Bradley through such current figures as Alasdair MacIntyre, Michael Sandel, Charles Taylor, and Michael Walzer, the proponents of the communitarian turn have left the idea of ethical community with little further filling, with the result that they have committed themselves to two fateful qualifications that ultimately undermine the solution they seek: 1) that ethical community is the exclusive structure of normative conduct and 2) that the concept of ethical community does not itself mandate the particular content of the roles, ends and association that it may take.

These qualifications of the idea of ethical community reflect an embrace of the type of thinking underlying Wittgenstein's private language argument. That argument took the inability of finding resources in the structure of consciousness for distinguishing between certainty and truth, or between subjective stipulation and objective validity, as indicative of how objectivity must be sought in given intersubjective conventions whose shared practices provide standards of justification that can be distinguished from merely subjective belief. Such conventions, however, cannot themselves be susceptible of any justification, since that would require appeal to some other measure of "objectivity", which is impossible if shared practices are the exclusive source of normativity. Accordingly, the intersubjective conventions that provide non-subjective meanings and knowledge are themselves contingent, ungrounded practices, with the same arbitrary lawfulness as games. Similarly seeking to overcome the limits of appealing to "the given" or to the form of individual willing as sources of normativity, communitarians have turned to ethical community as an historically contingent structure of shared practices with the same arbitrary content and the same exclusive norm-sustaining role as the language games to which Wittgensteinians turn as the locus of meaning and truth.

Thereby, on the one hand, communitarians absolutize ethical community, identifying all normative conduct with the activities ingredient in ethical community. Rendering the domain of ethics coextensive with the roles animating ethical association, this absolutization excludes any morality without community as well as any property entitlements that might be given independently of the distributive prerogatives enjoyed through membership in such putative ethical institutions as the family, the market and the state. Such denial of any normative action existing apart from ethical community rests upon the assumption that the only alternative is an atomistic ethic, grounding conduct in the acts of the self upon itself or upon external things and succumbing to the same pitfalls that undermine teleological and procedural theories. That there might be other forms of interaction involving rights and duties given independently of membership in some existing community is simply not considered.

Complementing this absolutization is a commitment to the formality of the concept of ethical community. At the same time that ethical community is advanced as the exclusive framework of normative conduct, the idea of that framework is held to leave undetermined what particular institutional configurations and correlative roles and aims it may involve. Since these details are not contained in the concept of ethical community, they are beyond a priori prescription. Instead, they must be left to the grasp of an historical understanding that describes the contingent conventions of history that alone provide content to the otherwise empty scaffold of the idea of ethical community.

This formality is of crucial significance for it directly resurrects the very same problems that the appeal to ethical community was designed to overcome. Just as teleological ethics could not demonstrate the universality of its particular candidate for the good, so communitarianism here renders the content enshrined in ethical community an arbitrary and transient phantom of history. And just as procedural ethics could not tie the universal form of willing that it privileged to any particular ends without falling back upon teleological appeals, so communitarianism has ended up leaving the universal idea of ethical community powerless to specify any particular institutions to concretize its unity. Although each putative ethical community may well comprise an institutional system reproduced by the roles its members constitutively perform, the historical contingency of its content renders its universality something wholly relative to the particular conditions of its existence. Far from being genuinely universal, the specific content of institutions and agency is only common to a particular, contingently given community, whose own existence is devoid of any further justification. Hence, it makes no sense to judge one community by the values of another. Moreover, the legitimacy of no community can be upheld from within, for all that any can lay claim to is the accidental fact of its own existence. Any successful transformation of one regime into another institutional order will thus have no more and no less rationale than the preservation of the status quo. Whatever community results will possess the same purely internal validity, whose fortuitous foundation makes its authority an arbitrary fate, which its members have no more reason to flee than to embrace. Although the roles they assume carry with them duties tied to the reproduction of the community to which they belong, once that membership be questioned, no ethical imperative remains. In effect, the communitarian turn offers little more than an injunction to occupy one's station. This might satisfy an Eichmann, but it otherwise reflects a formality in the idea of ethical community that rivals that of any categorical imperative.

No remedy is to be found by appealing to historical necessity and claiming that, although no ethical community is universal, each particular shape arises in a necessary sequence, leaving it the exclusive ethical option for its time. This gambit, offered by F.H. Bradley[1] and commonly misattributed to Hegel, only further weakens the ethical claims of communitarianism. To begin with, upholding the claim of any necessity in historical development is problematic enough. No empirical study of history can establish necessity, since not only is the accuracy, representative character and trustworthiness of all historical testimony subject to doubt, but all any study of historical fact can confirm are particular patterns of association of what has happened to have been observed. Alternately, any attempt to conceive a priori some necessary development in history is at pains to maintain the difference between history and nature upon which reference to anything historical depends. Yet even if historical necessity be granted, it can provide no remedy to the formalism afflicting the communitarian turn. For if the content of every ethical community is not only particular to an historical period, but a necessary product of historical development, how can the activities of its members possess the basic autonomy on which ethical responsibility depends? Far from being a regime in which duties and virtues have their life, ethical community would become a prisoner of necessity, resembling more the instinctual society of social insects than an association in recognition of any good. Instead of obtaining normative validity, an historically necessitated ethical community would be removed from the sphere of ethics and take its place as one more background condition beyond which the domain of right and wrong begins.

Liberating Ethical Community from its Communitarian Deformation

If the concept of ethical community is to provide a solution to the dilemmas plaguing contemporary ethics and serve as a guide to reconstructing modern institutions, the two defining dogmas of the communitarian turn must be supplanted. For this to be achieved, the concept of ethical community must undergo a dual transformation.

On the one hand, ethical community can no longer be conceived as a formal framework, whose content is relegated to historical accident. Instead, ethical community must obtain a universally valid content, bringing necessary concreteness to its concept. As the experience of teleological and procedural ethics testifies, such normative content cannot be supplied by appeal to privileged givens masquerading as highest goods or by recourse to privileged forms of willing posing as archimedean procedures of ethical construction. In both cases, factors are immediately advanced as sources of normativity that cannot be legitimated in accord with the very standard of legitimacy that they impose. Whether the legitimating standard be a privileged given content or a privileged determiner of sanctioned ends and institutions, the abiding distinction between legitimated conduct and legitimating factor, a distinction generic to foundational arguments, introduces this debilitating incoherence. What alone eliminates the distinction and the foundational dilemma it reflects is the identification of normativity with the reality of self-determination. Unlike any other shape of conduct, self-determination owes its measure to nothing but itself, overcoming the gap between legitimating factor and legitimated conduct. In this respect, self-determination alone escapes dependence upon extrinsic foundations and enjoys the self-grounding, presuppositionless character that fundamental normativity requires.[2] Accordingly, if ethical community is to escape the relativity of an externally conditioned content and take its place as a genuine ethical structure, it must be reconceived as an association of freedom, whose agencies, ends and institutions are defined by nothing other than their role as constituents in the reality of self-determination.

This does not signify that a normatively valid ethical community must somehow escape having natural preconditions altogether. Like any form of conduct and association, ethical community must share the natural features that underlie the existence of a plurality of agents. However, these prerequisites, comprising the physical, astronomical, chemical, biological and psychological conditions without which rational agents cannot interact at all, do not thereby prescribe what shape ethical community should have. Precisely because these natural conditions are the enabling prerequisites of all conduct, they cannot serve as principles for differentiating between the good and bad conduct that they equally make possible. Hence, the presence of such enabling factors in no way undercuts the normative independence of a community whose bonds consist in modes of freedom.

However, any community whose roles, ends and institutions are instead allowed to be prescribed by factors given independently of freedom will at best comprise a "natural" ethical community, with a tainted validity. Although the members of such a community, like the ancient Polis, perform roles in recognition of a common good that they reproduce as the existing framework of their activity, their stations reflect structures of domination, where hereditary rank, gender, race, cultural traditions and other heteronomous factors constrict autonomy. Insofar as the communitarian turn leaves ethical community determined by historical accident, it effectively reduces ethical association to a sphere of "natural" determination, where legitimacy is invested in given conventions devoid of any unconditioned validity, save for the purely contingent eventuality that history gives rise to institutions of freedom. In that case, which modern times have partially witnessed, normativity resides in the self-determined character of the conventions, rather than in their shared acceptance by those who practice them.

Reconstructing ethical community as an association of freedom might well appear to be at one with the task of determining self-determination that has preoccupied ethics and convulsed modernity ever since the claims of liberty supplanted the teleological appeal to a highest good. The great difficulty stymieing traditional attempts to conceive freedom has been their inability to conceive the free will such that both the form and content of its agency, that is, both who and what are willed, are determined by its volition. Lacking this double reflexivity, the will remains defined by factors it has not imposed upon itself, reducing its freedom to a merely formal capacity of choice. So long as the free will is conceived as a function of the self, it cannot help but suffer this limitation and have a given, natural form of agency that is the perennial precondition, rather than the product of its willing. In this respect, liberty, the faculty of choice, is a natural will, whose capacity to decide among given ends is indeed the presupposition of any act of volition. However, in the non-natural, historically emergent context of ethical community, individuals can very palpably exercise an artificial agency whose own character and ends are determined by the very willing that reproduces the association within which it alone can be exercised. In this way, for instance, individuals can exercise the specific political freedoms defining the agency of free citizen only as a member of a democratic state, whose institutions themselves consist in nothing but the political self-determination of its members. Hence, the character of their agency is itself the product of the institutional freedom they engage in, allowing them to exercise a mode of willing that is genuinely self-determining. One might therefore be tempted to consider normatively valid ethical community as identical with the reality of self-determination. Given the identity of freedom and normativity, one would then end up confirming the second dogma of communitarianism, which absolutizes ethical community as the exclusive framework of legitimate conduct.

However, the rehabilitation of ethical community cannot take this route and simply eliminate the formalist assumptions of communitarianism and supplant natural determination with the artificial freedoms of ethical association. The second communitarian dogma absolutizing ethical association as the exclusive mode of normativity must also be overturned. For, far from being coextensive with the reality of freedom, ethical community cannot coherently comprise a structure of self-determination at all unless two other modes of freedom are admitted whose autonomous agencies are not defined by membership in an existing community. These two types of freedom without community are the self-determinations constitutive of property ownership and moral accountability. Ethical community cannot enjoy normativity without recognition of their independent being because both property relations and morality play constitutive roles as necessary preconditions for exercising any freedom as member of an existing association.

Property and Morality as Preconditions of Ethical Community

Contrary to communitarian dogma, the two freedoms without community are not atomistic in character. Determining oneself as a property owner and as a moral subject both involve interacting with and in respect to other agents who do the same. To exercise the agency of a person, that is, to count as a free individual recognized to have entitled possession of at least one's own self, one must objectify one's will in some factor in a manner recognizable to and recognized by another person, whose own will enjoys the same recognized status. Otherwise, one simply has intercourse with things, exercising a technical mastery in which no question of entitlement can enter in. Similarly, to act morally, one must will and be held accountable for willing in respect to the right and welfare of other morally accountable agents. Otherwise, one's deeds have no moral dimension.[3] In each case, the freedom at issue involves an exercise of right tied to a corresponding duty to respect the correlative freedom of those who are bound to respect one's own like freedom. As such, the freedoms of persons and moral subjects irreducibly involve relations of plurality in which the exercise of right has an objectively recognized existence.

Yet, in engaging in these interactions, property owners and moral subjects still do not respect each other's entitlement on the basis of being members of any preexisting community. The rights and duties of personhood and moral accountability extend beyond the reaches of every particular institutional framework. It matters not whether one belongs to the same family, participates in the same society, or is the fellow citizen of another person and moral subject. In each and every institutional context, one remains bound to respect the due property entitlements and moral accountability of others and have one's own autonomy as person and moral subject respected. These rights are unavoidably basic because no valid institution of freedom can ignore them without undermining itself.

This is most apparent with respect to the right of persons to be respected as an owner. Exercising this right always involves being recognized as inalienable proprietor of one's own facticity, since otherwise one is liable to be owned by another will like any other rightless factor. Consequently, whenever respect for the right of persons is not incorporated in existing conventions, individuals are in the position of slaves, for whom no act is recognized as juridically their own. In such a situation, the individual can no more be held morally accountable than occupy any institutional role, such as family member, economic agent, legal subject or citizen, to which right and duty apply. Accordingly, no ethical community can give its members an autonomous agency, be it as codetermining spouse and parent, autonomous member of society, or self-governing citizen, unless they be recognized as persons owning at least their own body.[4]

The recognition of moral accountability is equally indispensable. If individuals are not entitled to be held responsible for acting with the right purposes and intentions and an understanding of what is good, how can they fulfill any roles as members of a community whose bonds consist in modes of freedom? The determinations of conscience may well conflict with the demands of rectitude required, for example, by citizenship, but if moral autonomy is totally ignored, rather than partially restrained, the possibility of ethical rectitude is itself undercut. Agents can hardly enjoy the rights and observe the duties of self-determined family membership, social responsibility, and citizenship if they are not held accountable for conscientiously acting with the right purposes and intentions. When moral autonomy is ignored, agents cannot determine the scope of their own deeds. Their purposes, intentions and conscience no longer have any bearing upon what they are held responsible for and how they in turn judge others. Such a situation is commonplace under conditions of "natural" ethical community, where slavery goes hand in hand with the submission of conscience to the external authority of given tradition, be it secular or sacred. Truncating the autonomy that can be exercised in community, this predicament deprives each role of all moral dimension, while leaving action at the margin of community, where one's station no longer determines what is to be done, determined by factors other than freedom.

Ethical Community as the Obligatory Complement of Property Right and Morality

Whereas ethical community can neither be nor be conceived without incorporating the independently determined freedoms of property and morality, these freedoms without community can be exercised prior to as well as spatially removed from ethical institutions. Nevertheless, property relations are no more able than moral strivings to guarantee the realization of their own freedoms by themselves. Neither can unequivocally determine nor securely uphold its own rights and duties in each and every case. If owners are left to their own devices, as defined by the constituents of property relations, nothing can prevent the outbreak of malicious and non-malicious violations of property entitlements for which no satisfactory solution is available. Whereas owners can always disagree in good faith in their interpretations of property boundaries and terms of contract, property relations contain no universally recognized authority to provide a binding adjudication of such disputes and award due compensation. Hence, unless owners come to an agreement to settle their dispute, non-malicious wrongs will persist. Similarly, the same absence of some higher respected power leaves malicious wrongs bereft of any authoritative charge, adjudication or punishment, leaving victims no other recourse than revenge, which risks being interpreted as a wrong of its own.

These limits to the security of person and property call for some remedy and, in the absence of an institutional solution, personal initiative is the only immediate option, an option characteristic of the moral situation, where the individual must independently determine the particular content of the good and bring it into existence through personal effort. However, recourse to moral agency only adds problems of its own.

The engagement in moral conduct does indeed actualize freedom to the degree that the limits of responsible action are recognized to be determined by the knowing and willing of the agent, in that moral subjects hold one another accountable only for that part of their deed and its consequences that are prefigured in their purposes, intentions, and conscience. Yet paying heed to the moral prerogative of subjectively determining the realization of right generates a dilemma brought to a head in each agent's effort to act on conscience. Here what purposes and intentions count as good is left to the independent determination of the agent, whose initiative is required to bring right into being. Yet, if others are to recognize the right of the moral agent to act on conscience, they are liable to fall into contradiction with the edicts of their own conscience. What one moral agent knows to be good based on conscience need not conform to what any others hold good. Since the good is not a merely private matter, but something objectively and universally valid, one cannot respect the conflicting conscience of another without denying the claims of one's own. The only escape from this dilemma on the basis of moral accountability is the fortuitous happenstance that one's conscience has spoken in unison with every other. Yet since such harmony depends upon purely external circumstance, it hardly upholds the autonomy of moral action.

The requirements for overcoming the limits of property relations and morality are therefore twofold. To overcome the limits of property, the reality of right must not consist simply in an external factor, outside of which willing occurs. Rather, the reality of right must incorporate the activity of willing that brings it into being. Alternately, to overcome the limits of morality, the subjective determination of the good must be such that what each agent independently aims at is inherently in consonance with the aims of others.

Ethical community, reconstructed as an institutional system of freedom, provides a solution on two levels.

On the one hand, since ethical freedoms have as their own prerequisite constituents the respected autonomy of person and moral subject, ethical community provides an existing institutional order in which property relations and moral accountability are objectively observed in conformity with the other community freedoms they make possible. Without incorporation within ethical community, property and morality either remain prey to incipient conflicts they cannot themselves resolve or get propped up by conventions given independently of freedom, thereby compromising their own validity. By contrast, within ethical community, the freedoms of owners and moral subjects get sustained on the basis of ethical self-determination. This does not involve a limitation upon ethical freedom, undermining its own autonomy. Because property and morality are themselves prerequisite ingredients in ethical freedom, upholding their self-determinations is part and parcel of ethical community's own reproduction.

On the other hand, ethical community overcomes the limits of property and moral self-determination by constituting a mode of freedom uniting their complementary dimensions. In ethical community, the good at which conduct aims contains the activity of its realization insofar as membership involves knowing and willing goals whose achievement reproduces the whole, thereby securing the unity of each member's volition with the performances of the rest. Consequently, the boundaries of the objectification of willing have a common and commonly recognized existence in the unity of the community, removing all incipient conflicts and equivocations so long as members uphold their roles. Moreover, since ethical institutions reproduce themselves in face of possible failures of rectitude, their constitutive activities must also involve overcoming deviations in conduct. In these ways, ethical community provides the objectively recognized realization of right that property relations fail to secure, as well as the universal conformity of subjective initiative that morality is ever seeking.

Hegel and the Rehabilitation of Ethical Community

With ethical freedom presupposing property and morality, and property and morality requiring incorporation in ethical community, the false dichotomy of atomism and ethical community must be supplanted by the threefold differentiation of property right, morality, and ethical community. Only then are the constitutive preconditions at hand for conceiving as well as realizing the institutions of freedom that comprise the complete system of ethical community.

The basic conceptual framework of this threefold differentiation has, of course, been pioneered by Hegel in his *Philosophy of Right*, which escapes the dual dogmas by which communitarianism deforms and delegitimizes ethical community. By conceiving right as the reality of freedom, and thereby identifying normativity with self-determination, and by then distinguishing the three fundamental modes of freedom as Abstract Right, Morality and Ethical Community, Hegel has provided the basic framework within which ethics can prescribe the different institutions of ethical association.

Yet when Hegel advances to differentiate ethical community as a politically self-regulated system of family, civil society and the state, he partially capitulates to the communitarian agenda, allowing all three ethical spheres to be branded by natural distinctions. Bending to the history of his and, alas, our own day, Hegel ties the family to a heterosexual monogamous union, where roles are defined by gender, dictating a privilege to represent the family in society and state for the husband and a bondage to domestic affairs for the wife. Whereas these curtailments of family freedom already limit social and political freedom by gender, Hegel makes both public spheres further hostage to natural privileges by giving monarchy and estates decisive roles in civil society and the state. Instead of enforcing equal economic opportunity among classes defined by different forms of market earning, Hegel retains estates, whose feudal distinctions define membership partially by hereditary rank and combine corporate and political privileges, undercutting the demarcation between civil society and state. By next making the head of state a hereditary monarch instructing an estate assembly, Hegel allows birth right to curtail the equal political freedoms of representative democracy.

Despite these major lapses, Hegel still provides a basic conceptual anatomy of family, social and political association that permits differentiating the modes of ethical freedom independently of extraneous factors. To begin with, domestic community can be identified as an immediate ethical association, whose unity is based on the shared feeling of its adult members to join together, irrespectively of gender differences and sexual orientation, into a common private sphere, for whose consolidated, but particular, right and welfare they are mutually responsible. Civil society, by contrast, can be characterized as a mediated ethical association, whose unity reproduces itself through the reciprocal pursuit of self-selected needs for what others have to offer as the instrument for achieving their own particular interests. Unlike domestic community, social community is universal in scope, comprising the exercise of a freedom extending beyond household and national boundaries alike, although the ends its members aim at remain particular in content. For this reason, the global community of interest in which civil society consists functions as the ground of interdependence whose appearance is the pursuit of personally selected interests constitutively tied to the satisfaction of the interests of others. Political association, for its part, has neither a simply immediate nor mediated unity. Its members are citizens not because of some shared feeling, nor as a result of aiming at particular ends distinct from political union. Rather, they exercise their constitutive political freedom by consciously willing the form of the whole body politic to which they belong. So engaging in acts of self-government whose aims are universal, citizens thereby reproduce the political institutions that preside over particular households and national society, allowing the totality of ethical life to form a self-ordered whole, resting on its own freedom. Achieving this self-determined unity, where what is mediated and what mediates are one and the same, political community unites the immediacy and mediation characterizing household and social community.

The necessity of this division of ethical community into distinct spheres of family, civil society and state is exhibited in respect to the scope of each association and the character of its defining ends. Reconstituted as an ethical community of self-determination, the family comprises an association particular in scope and particular in aim. By contrast, in civil society the association is universal in scope but particular in aim. Finally, in politics the association is universal in scope[5] and universal in aim when a state stands by itself, whereas the association is particular in scope and universal in aim, when one state interacts with others within the sphere of international relations. Thereby politics exhausts the logical possibilities of ethical association, certifying the completeness of the basic division of family, civil society and state.

With the conceptual resources underlying these rudimentary divisions, the system of ethical community can be duly reconstructed, purging all residue of illegitimate natural determination and developing the complete reality of the freedoms that should define membership in the household, society and the state.[6] Rescuing ethical community from the communitarian turn, it is this reconstruction that deserves to be the central agenda of ethics today, as much in theory as in practice.

*****

Although ethical community commands normativity by realizing a specific mode of self-determination, ethical community is subject to deformations whereby its roles, ends and association get defined by "natural" differences, given independently of freedom. The following investigation examines how such "natural ethical community" is afflicted by internal legitimation problems exposed by Hegel in his Phenomenology of Spirit. *Unveiling the intractability of these problems sets the stage for leaving behind both traditional community and communitarianism and exploring the foundation-free institutions of freedom.*

Notes

1 F.H. Bradley, *Ethical Studies* (New York: Oxford University Press, 1988), 190–192.

2 For a fuller treatment of how self-determination is the locus of normativity, see Richard Dien Winfield, *Reason and Justice* (Albany: State University of New York Press, 1988), Part III.

3 For a further discussion of this point, see "The Limits of Morality" in Richard Dien Winfield, *Overcoming Foundations: Studies in Systematic Philosophy* (New York: Columbia University Press, 1989) and "Morality Without Community" in Richard Dien Winfield, *Freedom and Modernity* (Albany: State University of New York Press, 1991).

4 For this reason, it makes no sense to argue that property entitlements presuppose distributive justice. Although the ethical associations of family, society and state may mandate adjustments in property relations in conformity with their own constitutive rights and duties, the normative claims of these distributive regulations all depend upon the prior recognition of personhood, without which participation in their associations is rendered impossible.

5 Admittedly, the scope of political association may be universal only relative to the particular boundaries of domestic territory. However, political association need not be bounded by a plurality of individual states for global political unity is not inconceivable.

6 For an attempt to reconstruct in detail the freedoms of civil society, see Richard Dien Winfield, *The Just Economy* (New York: Routledge, 1988), and Richard Dien Winfield, *Law in Civil Society* (Lawrence, KS: University Press of Kansas, 1995). For a reconstruction of family and state, see Richard Dien Winfield, *Reason and Justice* (New York: State University of New York Press, 1988), Part III.

Seven

The Immanent Critique of Natural Ethical Community

Despite recent philosophy's preoccupation with overcoming foundations in theory and practice, few anti-foundationalists have avoided enlisting given communal practices as the privileged ground from which meanings, knowledge standards, and ethical values get constructed. In so doing, they may have escaped the egological limitations of Cartesianism self-reflection, but they have retained the foundationalist assumption that justification must always operate in reference to predetermined parameters.

Hegel's *Phenomenology of Spirit* provides a timely tonic to these irresolute adventures in anti-foundationalism, for what Hegel presents in this work is an immanent critique of foundational justification in all its shapes. Most commentators have misconstrued the *Phenomenology* as a positive treatise providing a two-fold body of knowledge. On the one hand, the *Phenomenology* is commonly assumed to conceive the nature and interconnection of the different configurations of human culture. On the other hand, the work is alleged to traverse these shapes to attain an absolute standpoint achieving the unity of thought and being previously ascribed to the intuitive understanding of a divine intelligence whose thinking constitutes the objectivity it conceives, thereby surmounting the objectivity problem of traditional epistemology.

Such an interpretation is completely incompatible with Hegel's own account of the *Phenomenology*'s method and structure. In the work's introduction he plainly indicates that the *Phenomenology* engages in a pure observation of different shapes of consciousness whose endemic attempts at self-justification successively generate one another. Whatever knowledge claims arise are made by the shape of knowing under view and not by the phenomenological observer. If instead, the work were to make scientific claims about knowing, ethics, religion, cultural history or any other topic, the *Phenomenology* would illicitly take for granted the legitimacy of its own cognition and the character of the given.

The prevailing interpretations that treat the *Phenomenology* as a positive doctrine further contradict Hegel's repeated claims in the *Science of Logic* that the *Phenomenology* serves as an introduction to philosophy proper by eliminating the standpoint of consciousness as the paradigm of philosophical thought. Because philosophy must account within its own investigation for its method and subject matter or else fall prey to arbitrary assumptions, *Phenomenology* can hardly introduce philosophy's radically self-responsible discourse by being a positive doctrine. *Phenomenology* can only serve to introduce a foundation-free inquiry by playing a purely negative role, showing how attempts to do philosophy with presuppositions, that is, to validate normative claims by appeal to predetermined contents, undermine themselves on their own terms, paving the way for an alternative.

Hegel characterizes the topic under observation in *Phenomenology* as the opposition of consciousness or the form of representation. This structure is paradigmatic of foundational justification insofar as consciousness is described as a standpoint that justifies claims by appeal to a given standard of truth. In order to legitimate its claims, consciousness must establish the correspondence of what it takes to be valid (e.g. its putative knowledge, or its relation to what it putatively knows, the pole of for-itselfness) with the predetermined factor that serves as the criterion of its validity reflection (e.g. the pole of in-itselfness). Because, however, the required comparison involves knowing that criterion, the legitimating factor enters in only as it is represented, rather than as it putatively is determined apart from the legitimation underway. Consequently, so long as thinking remains foundational, operating under the opposition of consciousness by making claims whose legitimating standard is independently given, any self-critique ends up revealing that what serves as the validating factor is actually given only as it appears.

Significantly, the distinction between justificatory practice and its legitimating standard need not involve an ontological difference, where, for instance, the validating reflection comprises a purely mental domain opposed to external nature. Both sides of the opposition of foundational knowing may be ontologically alike, as is the case in pragmatism, ordinary language philosophy, and naturalized or historicized epistemology where cognition figures as a fixture of the world it addresses.[1] Moreover, the normative standpoint may be just as well practical as theoretical, as in the case of *Dasein*'s interest-laden care for the world in which it has its being. Further, the justificatory process may be intersubjective as well as egological. As the wealth of shapes in Hegel's *Phenomenology* indicates, all that is necessary for foundationalism to persist is that the candidate for validity be differentiated from the standard by which it is legitimated. As long as that distinction maintains itself, the form of representation, e.g. the opposition of consciousness, remains in force, with all its attendant pitfalls.

The *Phenomenology*'s observation of the shape of foundational justification in the section, "Ethical Action", has more to say about the internal difficulties of the communitarian turn than about an alleged historical emergence of self-consciousness, as one commentator has recently suggested.[2] After all, if history is distinguishable from nature by being the product of rational agency, self-consciousness is an omnipresent presupposition rather than an outcome of historical development. The communitarian turn does not make this blunder, but it does absolutize ethical community as the exclusive context for normative validity in conduct while reducing it to a formal structure, leaving historically contingent the particular roles, ends and institutions it involves. Hegel's analysis of Abstract Right and Morality in the *Philosophy of Right* makes clear why ethical community cannot be absolutized, showing how agents cannot exercise any rights of community membership unless they are already recognized as owners and moral subjects. Without that logically prior recognition, individuals are tantamount to slaves, rightless factors subject to the will of another.

On the other hand, the inherent dilemmas of communitarianism's historicization of ethical community are cast in relief by the course of "Ethical Action". That section of the *Phenomenology* provides this service by exhibiting the internal difficulties of an ethical standpoint that embraces an ethical community whose roles, ends and institutions are understood to have a given character. Although commentators are apt to treat "Ethical Action" as a description of Greek civilization, as filtered through the tragedies of Sophocles, what is observed is more properly specified as the ethical standpoint of a naturally determined ethical community, whose members occupy distinct roles based upon differences given by nature (e.g. gender), entailing specific duties whose pursuit reproduces the encompassing associations whose law is equally given. Although classical drama and Greek society at a certain time and place may well illustrate such a standpoint, certifying this exemplification is an historical interpretation falling outside the boundaries of phenomenological observation. What matters phenomenologically are not extraneous historical references, but the structure of the ethical standpoint and the course by which its validity reflection proceeds.

To begin with, the standard for conduct in a naturally determined ethical community consists in the preexisting communal bonds, which directly contain the correspondence of the substance of the community and the particular activities of its members, who reproduce it by following their prescribed roles. This incorporation within the normative standard of its correspondence with purportedly normative conduct exhibits the structure differentiating the shapes of foundational justification grouped under the heading "spirit" in contrast to those phenomenologically observed under the titles, "consciousness" and "self-consciousness". Whereas the standard of validity in the shapes of "consciousness" comprises something in-itself completely opposed to the domain of cognition, and the standard of validity in the shapes of "self-consciousness" has the form of knowing (that is, for-itselfness), in each shape of spirit, the pole of in-itselfness consists in the correspondence of in-itselfness and for-itselfness. In this respect, "ethical action", like every other shape of spirit, takes as its standard of validity the correspondence of in-itselfness and for-itselfness that the shapes of "reason" took as a starting point. However, in contrast to the shapes of "reason", which simply represent this correspondence as their fundamental norm, those of "spirit" exhibit the same correspondence in the pole of for-itselfness, that is, in the referring, knowing, or normative practice that measures its validity against the structure of correspondence it confronts.[3] In the case of "ethical action", the individual conduct that takes validity to reside in the given ethical community to which it belongs no less reproduces the integration of itself in that association with every fulfillment of its role.

Although the agent here does not cease to have personal motives and hesitations that may impede the fulfillment of duty,[4] none have any normative standing for the agent in their opposition to the execution of the roles mandated by membership in the preexisting community. What counts is the significance of action in the reproduction of the given ethical association, not the intentions that may or may not accompany it. Hence, when an Oedipus discovers how his action unknowingly violated the most vital tenets of household custom, his ignorance does not vitiate the crime for him or for other members of the community. Although the reflective standpoint of moral subjectivity may find paradoxical such retroactive attribution of guilt for actions done without full intent, that standpoint has no authority within the natural ethical community, and its objections can play no role in the validity reflection internal to "ethical action". Nor does Oedipus' experience of terror and remorse give him sufficient grounds to question the exclusion of conscious intention as the determining factor in assigning blame. After all, the pain of personal tragedy does not itself undermine the ethical fabric of the community, for which such misfortune counts simply as deserved, inevitable fruits of action. Something more must be at work, if the ethical standpoint of natural ethical community is to be its own undoing.

Hegel further characterizes the naturally determined ethical community as one differentiated into family and political spheres, governed respectively by equally given divine and human laws. Moreover, he characterizes ethical agency as differentiated into two types of roles, each focused upon upholding one of these sets of laws and accordingly expending its activities within either the family or the political arena. These two roles are not performed by the same individual, but allotted to different groups on the basis of the natural difference of gender, such that women are charged with family affairs and upholding divine law and men are guardians of political life and human law. These characterizations may well fit Sophocles' *Antigone* and actual historical features of ancient Greece, but it can be questioned whether the specific details are called for by the argument. Namely, an ethical standpoint embracing a naturally determined ethical community need not confront a community divided into family and political spheres governed by divine and human law, nor perform roles differentiated by gender. Consequently, if the breakdown of that standpoint depends upon the particular features of Hegel's depiction, where the competing demands of divine and human law entail tragic conflict, that breakdown loses its phenomenological necessity.

Although a naturally determined ethical community need not be differentiated into *oikos* and *polis*, nor be governed by distinct divine and human laws distributed to these spheres, nor allocate roles by gender, it will have a given order embodying a customary law upheld by individuals in occupying positions prescribed by given tradition on the basis of natural differences. Since the order must be determinate, it will have particular spheres, ruled by particular norms, and animated by particular roles. Moreover, the community in its entirety will be a particular whole, differentiated from other equally particular communities, subject to their own particular law. Due to the givenness of each ethical association, neither communities nor their internal spheres are already mediated by any pre-established harmony. Further, because members are immediately committed to the stations they occupy due to given custom and who they are by nature, incipient collisions cannot be removed by a conscious revision of roles and values. Any such conventional intervention would strike at the authority of the ethical community and give license to a disengaged revolutionary reflection. The same problem arises in relations between ethical communities, where the outbreak of war is always possible and no authority is available to mediate disputes on terms compatible with the self-understandings of different communal traditions. As a consequence, the ethical standpoint that acts in accord with a natural ethical community ends up unable to legitimate the particular content of its own association and its own role, both because it is merely given on a par with any other particular tradition and because it is liable to self-destruct in pursuit of upholding the particular sphere and particular community to which it belongs against other members and non-members alike.

Communitarianism implicitly accepts this dilemma as an inescapable fate, as if we were all condemned to escape tragic destruction only by being banal Eichmanns, clutching thoughtlessly to our stations, unable to find any ethical bearings beyond the shared practices to which we contingently adhere.

Hegel, by contrast, points to a way of redeeming ethical community in his *Philosophy of Right*: namely, by liberating it from historically contingent given tradition and reconstructing it as a system of institutions of freedom incorporating property and moral rights. Although Hegel's determination of the ethical community of family, civil society and state remains marred by holdovers of pre-modern traditions privileging birth and gender, he nevertheless identifies normativity with self-determination. This move heralds a departure from foundational justification in that self-determined conduct no longer seeks its legitimacy in antecedently given conditions. Instead, the difference between legitimating factor and legitimated practice is overcome, signaling in practice the same elimination of the opposition of consciousness that systematic logic achieves in comprising a self-developing system of categories, whose self-thinking thought takes neither method nor subject matter for granted.

Within the *Phenomenology*, however, the self-dissolution of a shape of foundational justification engenders a new shape taking as its standard of validity what the prior shape's standard proved itself to be within the validity reflection in which it figured. This inversion occurs due not to pragmatic considerations, but to how validity reflection must represent its standard to check whether its claims actually correspond.

Hegel does not follow "Ethical Action" with a description of the standpoint of moral reflection, as might suit contemporary Kantian opponents of communitarianism, but rather with a shape of foundational justification in which the ethical standpoint belongs to a community ascribing the formal right of property ownership to its members, who submit to an equally abstract rule extending over a domain unlimited by any particular traditions. Since the ruling authority recognizes nothing but the abstract personhood of its subjects, the particular content of its governance is left a matter of personal arbitrariness. The validity reflection internal to a natural ethical community can be seen to lead to this new configuration insofar as the given particulars of its roles, laws, and boundaries get sacrificed as the universal result of the incipient conflicts to which it is subject. The supersession of these given particulars is the truth that makes the exclusively universal dimension of an empire of property right a new standard for ethical justification.

By contrast, the emergence of a moral point of view will recover the dimension of particularity, not as given in existing custom, but as determined by the agent's conscience. But that is another story in the annals of the immanent critique of foundational justification.

*****

Whereas property right can figure as a foundation, ruling supreme in an empire where ethical community is sacrificed at the altar of private law, ownership has an elemental primacy as the basic component that every further freedom must incorporate. Although this incorporation does not make property right a foundation from which the other rights are derived, it does have implications for how property right is itself conceived. As the following chapter shows, central problems in the conception of property and contract can be resolved once the basic character of ownership is duly recognized. Doing so equally reveals how ethics must begin with property right to escape the dilemmas of foundational justification.

Notes

1 William Maker makes these points in *Philosophy without Foundations: Rethinking Hegel* (Albany: State University of New York Press, 1994), 60–61.

2 Susan Hahn, "Value Conflicts and Belief Revision in Hegel's *Phenomenology*", *CLIO*, Summer 1995, vol. 24, no. 4.

3 Robert Berman analyzes the structure of the different parts of the *Phenomenology of Spirit* in these terms in his *Categorial Justification* (Ph.D. Dissertation, New School for Social Research, 1983), Part II, Chapter 2.

4 Hahn suggests that Hegel thinks it is logically impossible for an agent (presumably the agent of "Ethical Action") to recognize an obligation, yet have countervailing motivations. The possibility of such "weakness of will" requires nothing more than the faculty of choice, which agency cannot fail to possess. See Hahn, "Value Conflicts and Belief Revision in Hegel's *Phenomenology*".

Eight

With What Must Ethics Begin? Reflections on Property and Contract

With What Must Ethics Begin?

With what must ethics begin? Traditionally this question has been largely ignored in deference to the more pressing issue of uncovering the ultimate principle grounding the legitimacy of conduct and institutions. How this principle is obtained has been of lesser concern than what it is. Such indifference to starting points is perfectly understandable, given the presumption that ethics must aim at unveiling a foundation whose given content qualifies it to prescribe what ought to be. Yet this search for a highest good, comprising the prior aim that conduct must realize to be just, has become discredited on the grounds that no purpose can have normative legitimacy owing to its given content alone. Whether it be an ideal of perfection dictated by reason or the primary end that all agents desire by nature, any such good falls prey to the skeptical challenge that questions whether any given content can legitimate itself and whether any object of desire can escape being a matter of fact rather than value, whose elevation to normative standing is only subjective and contingent.

With growing recognition of how the various candidates for highest good owe their veneration not to their content, but to the choice of whomever has ascribed their content its ethical primacy, attention has finally turned to the path that ethics follows in determining its subject matter. All too commonly, this attention has taken the form of an embrace of procedural ethics, which presumes that ethical norms are to be arrived at by being determined by some privileged determiner, whose process of determining provides the foundation supplanting the arbitrariness of teleological ends with the right method for sanctifying what is good. Whether this privileged determiner take the form of the choosing will, as in classical liberalism, or practical reason, as in Kantian ethics, or the contemporary siblings of the ideal speech situation of communicative ethics and the original position of Rawls, it always figures as something that, though not itself equivalent to what is normatively valid, provides the determining source of ethics. Whatever follows from this source's operation automatically commands obedience owing not to its content, but to its origin. Making what is good something derivative of the conditions enshrined in the privileged procedure may thereby well avoid the bare assertion of any ends that are at once duties. Yet it leaves unaccountable how the privileged procedure can justifiably be ascribed either its own particular description or its special role without invoking further legitimating principles in need of some other form of justification.[1]

In light of this difficulty, it is necessary to ask again, with what must ethics begin, recognizing that neither a highest good nor a privileged determining procedure will provide a starting point saving ethics from dogmatism. This question hardly seems to bear upon the theory of contract, yet it turns out that resolving the mysteries of contractual obligation equally resolves the problem of what starting point ethical theory must have.

Contract and property relations in general have been treated as secondary, derivative matters by both teleological and procedural ethics. Given the teleological strategy of first contemplating the highest good and then deriving from it the subsidiary value of the activities contributing to its realization, it is not surprising that both Plato and Aristotle view the legitimacy of property and contract as issuing from the distributive requirements of political association, whose supreme self-sufficiency and subordination of all other spheres renders it the preeminent realization of the highest good. Procedural ethics, despite repudiating the primacy of the good over right, has equally treated property entitlements and contractual obligations as matters without any indwelling justification, deriving their right instead in function of the social contract, the application of practical reason, the deliberations of an ideal community, or some other determining factor that legitimates a certain regime of property and contract in view of its own prior concerns.

In both cases, as might be expected, the derivative legitimacy of property relations is accompanied by the relegation of corrective justice to something whose own validity depends upon the prior realization of distributive justice. Given, following Aristotle, that the proportional principles of distributive justice have sway as governing rules for securing the preeminent goodness of political association, the arithmetic principles of corrective justice can have authority only if property has already been allotted according to merit, allowing those who are most virtuous the best opportunity to realize goodness. By the same token, if some privileged procedure is to mandate a fair distribution of primary goods so as to promote the equal liberty of individuals, then the rules of corrective justice will have their legitimate application only in conformity with a preceding achievement of distributive justice. As each case makes manifest, so long as property entitlements are conditioned by distributive requirements, corrective justice, concerned as it is with fairness in transactions involving given property, will equally depend upon the dictates of distributive justice.

Hence, far from raising the question of with what must ethics begin, the problem of property entitlements and contractual obligation seems only to enter when ethics is well under way, provided, of course, that one accepts teleological and procedural approaches as the only options available.

Yet as Peter Benson has shown with great clarity and conceptual rigor,[2] contractual obligation and, by implication, property entitlements in general cannot be coherently conceived if distributive concerns are allowed to underlie them. Broadly speaking, all such attempts either make the fatal mistake of permitting entitled acts of will to be grounded on the subjective and contingent factor of inclination or commit the *petitio principii* of introducing distributive requirements as determiners of property entitlements, when the claims of distributive justice transcend inclination only by incorporating entitled ownership as part of their own basis. This signifies, by extension, that corrective justice cannot be made intelligible as an adjunct to distributive justice. [1173]

What option are we then left with? If property entitlements and contractual obligations are not grounded on antecedent distributive concerns, can they be accounted for independently? In turning to Hegel's conception of Abstract Right as a resource for exploring this option, Benson attempts to show that the right of property at hand in contract is nonderivative and radically rudimentary, comprising the elementary mode of self-determination that all other forms of right must presuppose and incorporate. In so solving the mystery of contractual obligation, Benson effectively answers the question, with what must ethics begin? Abstract Right is the only tenable starting point, and only when it is adopted without appeal to any other normative principles, can property relations be made intelligible.

Unfortunately, Benson is a shade timid in proposing his dual solution. Unlike Hegel, Benson restricts himself to the claim that, under the *assumption* that autonomy be the principle of ethics, Abstract Right is where we must begin, first in making sense of property entitlements and contractual obligations, and then in accounting for the other institutions of right. This suggests that autonomy itself can comprise a first principle determinative of, yet distinct from all ethical structures, including property relations. Such is the strategy followed by liberal theory and Kantian ethics, which both treat freedom as a first principle of ethics, thereby embracing a form of foundationalism where freedom gets construed as a privileged determiner dictating modes of conduct and institutions. This foundational interpretation of freedom contradicts the logic of self-determination. If freedom is treated as a principle conferring legitimacy upon practice by determining it as a derivative reality, freedom relinquishes its own autonomy by determining what is other than itself, whereas the derivative reality it enshrines is equally unfree, being determined by an antecedently given standard. This outcome exhibits the perennial pitfall of any appeal to foundations: what is ascribed normative legitimacy remains distinct from the factor that is supposed to confer validity, so that the primacy and description of the legitimating factor cannot itself enjoy normative validity, which is defined not by being the foundation, but by being determined by it.

The decisive insight of Hegel's philosophy of right is that the dilemmas of foundationalism can only be overcome if the conduct and institutions to which normativity is ascribed are self-determined, for only what is self-determined owes its character to no external givens and can exhibit unconditioned universality. In that case, the structures of justice need not owe their legitimacy to factors that are not just themselves. Instead, the reality of justice will be at one with the reality of self-determination, whereby action will not only be free of determination by subjective and contingent factors such as inclination, but by any terms not ingredient in the structures of freedom themselves. Thereby ethics liberates itself from the foundational appeals to teleological ends and privileged procedures to become a philosophy of right, insofar as right comprises the realized system of self-determination. The starting point of ethics will therefore consist not in unveiling a first principle of the good or a method of deliberation for generating right outcomes, but rather in identifying the minimal reality of self-determination, on which all further freedoms depend. It serves not as their determining standard, but as a structural component that they cannot fail to incorporate in their own irreducible spheres. If this minimal reality were instead treated as a criterion for legitimating and deducing the character of the other institutions of right, it would only reintroduce the ethical foundationalism that the philosophy of right can otherwise overcome.[3]

What then is this minimal reality of freedom to be? Benson, taking his cue from Hegel, claims that it is personality, the generic agency exercising the mode of freedom constitutive of property relations. Understood as the form of agency specific to the interaction of property relations, personality can be neither a natural capacity, such as the choosing will all agents possess as the given condition of all their conventional involvements, nor a capacity of the self, defined independently of relations between agents. Yet, Benson characterizes personality as simply the capacity of the agent to act independently of determination by desire. [1160] Can this be either the characteristic agency of parties to contract or the form of autonomy with which ethics must begin?

Personality as the Starting Point of Right

If personality is to provide the starting point of right, it must comprise the minimal realization of self-determined agency, presupposing no other modes of freedom, yet providing a structural component ingredient in all further spheres of right. Can it then be defined, as Benson suggests, as the mere capacity of the will to be free of all determination by inclination? That universal dimension of the will, comprising the negative freedom of being unbeholden to any particular ends of desire, may indeed be a precondition of rational autonomy, as Kant so emphasized. But so, one could add, is the corporeal being of the agent, desire itself, and every other factor without which rational agency cannot operate. The preconditions of rational autonomy are one thing, the minimal reality of self-determination is another. Only the latter comprises a bona fide topic of the philosophy of right.

From Hegel's point of view, the fatal pitfall of the Kantian conception of autonomy is that it fails to advance beyond the formal capacity to will independently of inclination, remaining caught with a truncated vision of freedom, one which may exhibit universality, but fails to achieve a particular determination in which the universality of the will remains preserved. As Benson himself emphasizes, the will, no matter how minimally realized, is only free if it succeeds in individuating itself independently of any given ends. This requires that the agency it exercises be one that not only determines its own ends but the very form of its own activity. Yet, to be able to will both requires that the willing individual engage in an interaction with other agents such that they all succeed in individuating themselves in terms of the interrelationship they each concomitantly will into being. Then, both the ends they will and the form of agency they exercise can be specific to an interaction that is itself nothing but the realization of their interrelated self-determinations. Otherwise, if an agent merely acts in reference to things or in relation to itself, the form of its agency always precedes its activity, as the natural condition of the choosing self. Hence, technical and otherwise monological action can never establish an artificial mode of autonomy whose form and content are equally products of willing. Such, however, is precisely what is at hand in the enacted conventions of property relations, moral interrelations, family association, civil society or politics, once these are reconstructed as institutions of freedom.[4]

Consequently, if personality is to be the most elementary mode of self-determined agency, not only must it constitutively will ends that maintain its freedom from inclination, but it must irreducibly involve the framework of a plurality of persons, where the form and content of their actions is predicated upon their interrelationship. On these terms, the abstractness of personality lies not in its reduction to the negative liberty of the choosing will, but in the fact that the particularity of its self-determination lies outside the agent in an immediately given thing. As Benson himself points out, what allows an external factor to provide the particular content by means of which personality determines itself is that this factor not be appropriated as an object of desire, chosen for the sake of its own natural features. [1169] Instead, the external factor rather operates as a mere receptacle of the person's free realization as owner, appropriated for the sake of giving personality the particular determinacy by which owners succeed in individuating themselves from one another solely in function of their willing, acquiring a particular objective domain embodying the personal autonomy of each. [1172]

What makes this mode of self-determination irreducibly elementary are two complementary features. First, the particular objectivity that realizes the will's self-determination is completely immediate, comprising a purely given entity, with no further qualification beyond its being other than a will. All additional relations of right do involve objectifications of the will, but mediated by such further factors as the intentions of the agent, the particular right and welfare of the household to which the agent belongs, the need for commodities of the individual, the proceedings of legality, and the all-encompassing concerns of self-government. Conversely, the agency that individuates itself in such a purely immediate objectivity is itself immediate in that its constitutive autonomy involves determining itself as no more than a person whose will has a particular embodiment recognized by others. That is just what minimally comprises self-determined agency. By contrast, every other relation of right will involve an agency whose self-determination consists in objectifying itself not just with no further qualification, but doing so in respect of its morally significant purpose and intention, its membership in a family, its satisfaction of market needs, and its subjection to and self-rule by law. For this reason, it is a category mistake to suggest, as Benson does in calling personality a free and equal moral agency, [1088, 1149, 1150] that the immediate agency of property owners is equivalent to the reflected agency of moral subjects, who interact not simply in terms of objective embodiments of their respective wills, but in regard to the purpose and intention underlying their deeds.[5]

Being wholly abstract in the above manner, personality certainly involves freedom from determination by inclination. However, it no less essentially contains the positive feature of being an owner, namely, being an agent whose specific identity is due to the exercise of will that has given itself an external embodiment concording with that of others. Although this is true of all property relations, being an owner has its own minimal determination that is distinct from the other derivative modes of ownership that presuppose and incorporate it. As Hegel, and Benson following him, point out, this minimal realization of ownership is wedded to a particular objectification on which all further objectifications depend. Namely, to determine themselves as owners, individuals must first lay claim to their own bodily existence on whose inalienable ownership depends every other appropriation and use of property.

Original Appropriation

No things can be taken into ownership unless the agent has ownership of its own body, for otherwise, the agent is as good as a slave, incapable of exercising any recognized independence from the will of others. Yet can ownership of one's own corporeal reality be something a person enjoys immediately, without any act of will, as Benson maintains? [1178] If ownership is the embodiment of the will in a thing, providing it with a domain exhibiting the self-determination of the agent, how can this be achieved in the body when its possession by the agent is not mediated by that agent's will? As a person, the agent is self-determined, which is to say, that what the agent is is something the agent has willed.

Moreover, can either the agent's ownership of its own body or its appropriation of an unowned thing be understood as a relation of personality and a thing, independently of an interaction among agents? To have entitled dominion over one's body automatically refers to other equally entitled persons insofar as right always involves mutual recognition among a plurality of individuals, who each enjoy their right only in conjunction with being duty-bound to respect the correlative right of others.[6] Hence, not only is ownership of one's body mediated by an act of will, but mediated by similar acts of will of other persons who engage in the reciprocal practice of honoring each other's exclusive right to their own bodies. As with every structure of right, it need not precede in time the other property relations it logically precedes. However, it must be at hand in conjunction with their operation if such relations as the recognized appropriation of other objects and the entering of contract are to occur in any coherent fashion.

Contract

Given the normativity of self-determination and how self-determination consists in structures of interaction in which participants determine both what they will and the form of artificial agency they exercise, the mystery of contractual obligation is ripe for solution. Far from requiring any appeal to prior principles, such as distributive imperatives, the legitimacy of contract rests on nothing more than the demonstration that contract is itself an interaction consisting in an exercise of self-determination on the part of its participants. The concept of Abstract Right provides the resources for this demonstration by showing how property relations are structures of freedom, presupposing no other relations of right, yet basic for all further normative practice. Once personality has been determined in conjunction with the original appropriation of property, we have at hand all that is needed to comprehend how contract operates as a structure of right. The plurality of property owners has been accounted for, together with the right of alienation of property inherent in ownership[7] and the basic acts of will by which ownership is established. Now, all that need occur is that owners simply choose to transfer property to one another in respect of each other's entitlements as owner. On this basis, as Benson points out, the traditional principle of no liability for nonfeasance is confirmed, [1082–1083] for the parties to contract are free of any obligations to enter into contract with others in function of their desires and welfare, just as whatever contract they do enter has legitimacy solely in respect of their property entitlements, regardless of whatever other expectations accompany their transaction. If need and welfare are to figure as matters of right, it will have to be in view of further modes of self-determination, transcending, yet incorporating property relations.[8]

Contract, so understood as simply the transfer of property from one person to another can, as Hegel maintains, occur in two basic forms: as a gift contract, where one person alienates property to another who agrees to take ownership of it, and as exchange, where one person alienates property to another and receives property that the other has alienated in return. Benson explicates their respective processes in a close exegesis of Hegel's argument. Yet, on several points, Benson introduces claims of his own that either illicitly restrict the argument or obscure its moves. Namely, Benson maintains that contract occurs only between two persons, [1176] that the continuity of contract during the transfer of property rests upon the parties taking into possession the same thing, [1185] and finally that contract entails a juridical principle of equivalence that can be applied in terms of the market price of the goods that are exchanged. [1195]

To begin with, need contract occur only between two persons? Property may be an entitled possession that is exclusive, yet this does not prevent the owner of some thing from being a plurality of individuals, who have entered an agreement to become joint owners. In this manner corporate as well as cooperative ownership can arise in function of abstract right without in any way contradicting the elementary form of private ownership. Although Hegel does not explicitly refer to such joint ownership in his discussion of Abstract Right, it provides the conceptual basis for the joint property of the family,[9] as well as for the various forms of joint ownership that figure within the commodity relations of civil society.

By the same token, contract may occur not only between two parties, be they individual persons or corporate persons representing the common ownership of a plurality, but among three or more separate participants, who all agree to transfer the use and ownership of their respective properties. Contract, minimally conceived and minimally realized, does require only two parties, and what pertains to the form of their agreement does characterize the basic relation of any contractual agreement. Yet this does not prevent the concept of contract from being exemplified in agreements involving transfers of property among a larger plurality, such as figure so prominently in the regulated market economy of civil society.

More significant, perhaps, is Benson's claim that in contracts of exchange there is a juridical requirement of equivalence that uses market price as its standard. Although, as Benson himself shows, contract provides the conceptual resources for determining the value of exchanged property, it does not, and indeed, cannot, from a systematic point of view, rely on the economic category of market price in determining this value. Market price is intelligible only in terms of the specific interaction of commodity relations, which do indeed incorporate property relations and contract, but within the civil context of interdependent need where property can figure as a commodity, embodying not simply the will of the person, but the freely determined, yet complementary need of market participants. Thus, contra Marx and all other purveyors of economic determinism, whereas market relations presuppose abstract right as a structural component, contracts of exchange cannot owe their character to the more complicated interaction of economic freedom.

The juridical requirement of equivalent exchange in real contract is a simple matter, which Benson's account of the common will underlying exchange contract tends to obscure. According to Benson, when parties to an exchange enter into their contract, they each become simultaneous owners of the same property. Otherwise, the continuity of the transaction would be broken, and the mutual transfer of property could not occur. [1185] Yet does the requirement that one party's alienation of property be at the same time its appropriation by the other, a requirement that makes possible a distinction between the agreement and performance of contract, entail that both parties momentarily own the same property? What the common will of contractual agreement determines is not such momentary common ownership, but rather the instantaneous transfer of title of each party's property. This jointly willed exchange of title occurs immediately upon whatever formalities comprise the signing of the contract, in distinction from the temporally and spatially differentiated acts of delivery in which consists the performance, as opposed to the agreement, of the contract. The moment title is transferred, the exchanged properties are set in a relation of equivalent exchangeability or exchange value. Although each party receives a different property, each thereby receives something of equal exchange value. The equivalent exchange value exhibited in each might thus be loosely called the common property each party simultaneously acquires, [1192] provided it be understood that the equal exchange value comprises the common relatedness of two discrete objects. This determination of exchange value is entirely internal to the transaction, resting upon nothing more than the particular terms that have been agreed upon, terms that stipulate which properties are exchangeable with one another within this particular transaction. What exchangeability the properties will enjoy in future transactions depends entirely, at least within the framework of abstract right, upon whatever specific contractual agreements happen to be reached. No matter how the terms may vary, so long as the parties enter contractual agreement in function of honoring each other's prerogatives as owners without coercion or misrepresentation, the equivalence of exchange value of the exchanged properties cannot help but be observed, embodying as it does the concordant free resolve of the contracting parties.

These minor revisions in conceiving how contract involves equivalent exchange in no way undermine the important conclusion Benson draws from Hegel's analysis: that contract entails its own nondistributive principle of fairness, posited in legal form in the common law doctrine of unconscionability, according to which contract enforcement should take into account whether the considerations exchanged be adequate. [1090] Hegel's discussion of fraud highlights how this principle is supported by abstract right independently of any external references to welfare, need, or market price. Although the prerogatives of alienating one's property leave parties to contract entitled to exchange their property at whatever terms they choose, establishing a purely conventional exchange value for the properties they transfer, they are still under mutual obligation to make clear to one another the true condition and character of the property they are offering so that what each agrees to receive accords with what they actually obtain. This imperative rests upon no extrinsic distributive concerns, but simply upon the principle inherent in contract that the parties to the agreement will in conformity with one another.

Nevertheless, when the administration of law gives abstract right a publicly enacted reality, it does so in reference to the further modes of self-determination constitutive of civil society,[10] wherein the economic welfare of civilians becomes a matter of right, exhibited in the principle of equal economic opportunity. For this reason, determining how law should regulate property and contract is not something to be done solely on the basis of abstract right. Still, without abstract right understood in its own terms, neither the institutions of civil freedom, nor the rights of persons can be made intelligible. Benson's rehabilitation of Hegel's doctrine of Abstract Right is therefore a necessary step towards the rational reconstruction of law's empire.

*****

The relationship between property rights and the other institutions of freedom comes to the fore in the tensions that seem to challenge the unity of the common law. As the next chapter shows, the apparent conflicts between right and welfare can be defused through proper consideration of the concrete connections between property and the different spheres of ethical community.

Notes

1 For an extended critique of both teleological and procedural ethics, see Richard Dien Winfield, *Reason and Justice* (Albany: State University of New York Press, 1988), Chapters 2, 4, and 5.

2 See Peter Benson, "Abstract Right and The Possibility of A Nondistributive Conception of Contract: Hegel and Contemporary Contract Theory" (*Cardozo Law Review*, Vol. 10, Mar.\Apr. 1989, Nos. 5–6). All numbers in brackets refer to page numbers of this essay.

3 For a more detailed exposition of this argument, see Richard Dien Winfield, *Reason and Justice*, Chapter 8.

4 This argument is developed at greater length in Richard Dien Winfield, *Reason and Justice*, pp. 158–165.

5 For a detailed account of moral autonomy in its distinction from personality and the agencies specific to ethical community, an account reconstructing Hegel's analysis of morality in his *Philosophy of Right*, see Chapter 8, "The Limits of Morality", of Richard Dien Winfield's *Overcoming Foundations* (New York: Columbia University Press, 1989).

6 Benson does recognize that each person's right over his or her bodily existence exists only in relation to others [1179], but Benson fails to emphasize how this relation is an interaction, consisting in interrelated acts of will.

7 Although ownership must always be or accompany inalienable ownership of the owner's bodily existence, that primary ownership makes possible the alienability of all other property, as well as the formation of consolidated ownerships, be they familial, corporate, or public, in which persons form associations that can acquire property of their own.

8 In *The Just Economy* (New York: Routledge, 1988), I have explored at length how need and welfare do figure as matters of right in conformity with the normativity of freedom,

namely, as factors of the specific mode of self-determination characterizing publicly regulated markets. As I there argue, following Hegel in his analysis of Civil Society, whereas the structures of economic self-determination necessarily incorporate property relations, the normative claims of economic welfare can only be coherently conceived independently of any procedural analysis of distributive justice.

9 This is not to say that the family is reducible to a contractual relation, but only that one aspect of the household, that involving jointly held property, incorporates ownership and agreement, albeit with peculiarities owing to the context in which it is embedded.

10 Benson touches upon this on pp. 1151ff.

Nine

Unity in the Common Law?

A specter of disunity haunts the common law, threatening to throw property, contract, tort, and penal law into a crisis, where competing paradigms stand intransigently opposed, undermining any claims of coherence and giving sober proof that legality is a battleground of equally unjustifiable ideologies where only force wins out at the end. The impending crisis pits advocates of liberalism, affirming the primacy of the right over the good, against communitarians, upholding the priority of commonly shared ends embodied in an historically given community. Yet although the conflict parallels what many take to be the exhaustive options of ethical thought, the difficulty extends beyond theoretical dispute into the actual practice of common law, where at every turn, tendencies promoting welfare clash with tendencies upholding the formal right of ownership. In face of such division in both theory and practice, the dangers of idealism seem hardly surmountable by fidelity to law or by reflective equilibrium, for if disunity pervades legal thought and convention, neither appeal to the given can locate a coherent kernel in the conflicted shell.

Alan Brudner offers an escape from this impasse in *The Unity of the Common Law: Studies in Hegelian Jurisprudence.*[1] As his title announces, Brudner seeks to overcome the disunity in the common law by drawing inspiration from Hegel. In Hegel's concept of spirit (*Geist*) Brudner purports to find the resources for comprehending how the competing tendencies in common law are actually components of an encompassing whole within which each seemingly opposing element affirms its counterpart as a necessary precondition of its own realization. Brudner identifies the unifying foundation as *dialogic community* and claims that it provides the basis of legality enabling the common law to retain coherence and reconcile the formal right of person and property with the welfare considerations of the common good. Moreover, dialogic community allegedly provides this service without appeal to any external standards; rather, the competing tendencies in the common law attest to its underlying truth by virtue of how their own efforts to assert exclusive hegemony end up demonstrating their mutual dependence.

Brudner's turn to Hegel arises from the prior jolt of Critical Legal Studies' denial of a unity in common law, a denial unmasking the ideological character of the competing legal paradigms and thereby effacing the divide between law and partisan politics. [290] Critical Legal Studies thereby awakes us from our dogmatic slumber, where opposing proponents of the right and the good assumed the exclusive validity of a single option. However, instead of embracing the nihilism at which Critical Legal Studies arrives, Brudner takes Critical Legal Studies' legal deconstruction as an epitome of contemporary common law consciousness, testifying to the incoherence of all one-sided groundings of law. [290] Whereas Critical

Legal Studies sees no other alternative, Brudner sets out to demonstrate that the absolutization of one-sided ideology is itself a dogmatic crutch that collapses under its own weight.

Brudner's investigation accordingly advances in three stages. First, he lays out the two opposing tendencies in each sphere of the common law. Then, he attempts to show how each tendency, developed on its own terms, undermines itself when it is independently advanced. Finally, he endeavors to demonstrate that the two opposing tendencies require one another for their own maintenance, and that this is achieved within a framework of dialogical community, which proves itself to be the foundation and unity of the common law by integrating the formal liberty of personhood with the welfare demands of the common good. Brudner draws upon Hegel at each stage of his analysis. In identifying the contrasting elements of the common law, Brudner relies upon interpretations of Hegel's accounts of property right (which Hegel calls "Abstract Right") and morality, respectively. In unveiling the one-sidedness of each element, Brudner adapts Hegel's own account of the limitations of property relations and moral conduct. Finally, in developing dialogical community as the encompassing unity of the two paradigms of the common law, Brudner draws upon Hegel's concept of Spirit (*Geist*), as presented in the famous master-slave analysis of Chapter IV of the *Phenomenology of Spirit*. The result is a doctrine of law in which formal right and community can coexist, where trespass and negligence can retain their respective spheres, where tort litigation can proceed in conjunction with legislated social insurance, and where penal justice can mete out retribution while attending to welfare.

Given Brudner's strategy, several questions come to the fore. First, has Brudner accurately identified the key elements of the common law and are these elements conceivable in terms of Hegel's categories of abstract right and morality? Secondly, do the opposing tendencies of common law defeat themselves when independently advanced and do their alleged limitations correspond to the limits Hegel ascribes to property relations and moral autonomy? Thirdly, does dialogical community coherently unify the elements of common law and can it be construed on the model of Hegel's concept of Spirit? Finally, can the independent self-annulment of the two paradigms of common law and their integration in dialogical community sanction their normative validity as indispensable components of legality?

The answers to all of these questions end up turning upon a striking disanalogy between Brudner's application of Hegelian categories of property, morality and spirit and Hegel's own treatment in the *Philosophy of Right*. Brudner begins by distinguishing the concurrent poles of the common law as the legal paradigm of formal right, in which person and property are upheld in indifference to welfare, and the legal paradigm of concrete effective freedom, where welfare concerns take center stage in conjunction with promotion of the common good. These two strands are associated with liberal theory and communitarianism, respectively, to the degree that liberalism affirms the primacy of right over the good, while communitarianism enshrines the commonly recognized values of members of an historically given community as the locus of ethical normativity. Brudner, however, makes the further association of identifying formal right with Hegel's concept of abstract right and of identifying concrete effective freedom with Hegel's concept of morality.

The association of formal right with abstract right might seem benign enough given how Hegel emphasizes that persons determine themselves as owners simply by embodying their will in external factors without regard for welfare, motive, or any other consideration distinct from giving their volition a recognized domain. Indeed, Brudner deflects one possible source of confusion by noting that although Kant's conception of property closely resembles that of Hegel, Kant identifies right with formal right, whereas Hegel regards abstract right as only the most elementary member of a series of more comprehensive structures of right, which come to include the supervening practices of moral, family, social, and political freedom. [166] Nevertheless, Brudner begins to muddy the water by associating formal right with the negative freedom of an atomistic agency to choose among given alternatives and by emphasizing how property relations that precede contract, namely, original acquisition and use, involve the solitary volition of the person without regard for the volitions of others. In this respect, Brudner emphasizes that the negative character of duty in property, where obligation consists in prohibitions against violating the property of others rather than positive duties of care and concern for their interests, reflects that atomistic character of property right. Here Brudner brings in Hegel, claiming that Hegel's account of original acquisition and use of property proceeds monologically, as a relation of the self to things. Hegel's own text somewhat justifies his interpretation, since Hegel tends to downplay any accompanying relation between persons. Nevertheless, as Brudner himself admits, [58] Hegel's account of how acquisition must take a recognizable form equally implies that any acquisition of property must occur in relation to the recognition of other persons, whose own status as owners rests upon the same recognition process. This role of recognition is absolutely crucial for property, since without the relation between owners the very notion of right disappears, for right involves a universal relation of volitions, where agents recognize one another as each having an entitled sphere of choice that all others are obliged to respect. Without a contrastive relation to the embodied wills of other persons, an individual can never establish more than physical possession of any object. Consequently, property, like every other type of right, cannot be an atomistic relation of an agent and things, but must be an intersubjective relation between agents, albeit by way of external factors. That property right involves prohibitions against violating the ownership of others rather than positive duties of care reflects not the absence of recognition, but rather the abstraction from interest as a factor in the constitutive self-determination of owners. To be an owner involves simply that one's will be recognized to be embodied in an external factor. Whether that factor has any further relation to an owner's needs, desires, or motives, or any relation to the needs, desires or motives of others may have relevance for further structures of right, where different types of welfare play a role, but not to the consideration of whether it is someone's property. Brudner is certainly correct to associate Hegel's concept of abstract right with the abstraction from interest and the restriction to the formal considerations of person and property, where right entails negative prohibitions against violation of ownership. Where Brudner goes wrong, however, is in associating property right (and Hegel's account thereof) with an atomistic relation, an association that comes to play a key role both in Brudner's analysis of the self-contradiction of formal right and in Brudner's depiction of dialogical community.

Brudner introduces a complementary ambiguity in how he describes the second pole of the common law, where concerns of welfare and the good allegedly challenge the abstract limit of formal right. He identifies this alternate legal paradigm as a doctrine upholding the conditions of effective self-determination, where individuals exercise the right of having their welfare secured as an element in the promotion of the common good, a good comprising the attainment of the deliberate, coherent self-selected life-plans of individuals. Here the respect for person and property gets curtailed by a respect for the interests of individuals as defined by the requirements of the chosen life in which their self-determination is concretely expressed. Brudner subsumes the resulting rights under the heading of "rights of intention", directly associating these with the rights that Hegel attributes to the moral subject in the analysis of purpose and responsibility, intention and welfare, and conscience and the good comprising the section on "Morality" in the *Philosophy of Right*. [37] Brudner maintains that such common law fixtures as proprietary estoppel, "quasi-property", contractual licenses, and constructive trust embody the judicial concern for effective autonomy that the rights of intention make obligatory. [37] These rights protect individuals from the legal power of consequences beyond their rational control and of interests exclusive of their own. [37] Brudner describes the effective autonomy that such rights uphold in terms that reflect Kant's account of morality: the positive freedom such autonomy comprises is defined in contrast to the heteronomy of being a means to another's plan, of acting from given motives not integrated into a reflected plan of life, or of otherwise being subject to external forces. [68] The connection to Kant is particularly appropriate since Kant has no other positive freedom to offer in distinction to the formal freedom of right than the autonomy of moral conduct.

With Hegel the situation is entirely different. Hegel does regard morality as a sphere in which subjects determine their conduct towards one another in reflection of their purposes and intentions, thereby pursuing interests whose fulfillment represents a welfare whose goodness becomes an issue for conscience. Yet, Hegel hardly limits considerations of welfare to moral reflection. Instead, he extends "positive freedom" to the institutional domains of family, civil society and state, each of which introduces sanctioned ends whose fulfillment extends beyond the upholding of property rights and requires limitations upon ownership. Consequently, if the common law embodies restraints upon ownership that reflect a concern for welfare, it would make sense to refer to the normative requirements of family, social and political freedom, whose very concrete determinations contribute a definite content sorely lacking in Brudner's notion of an effective freedom to realize a well-considered personal plan of life.

Brudner's problem, however, is not simply a matter of neglecting to connect welfare considerations in the common law to the ethical freedoms of family, civil society and the state. His immediate difficulty lies in drawing out any legally relevant welfare considerations from the rights of moral accountability. His appeal to Hegel's analysis of morality brings this problem to head, for if anything, Hegel's discussion suggests that moral autonomy is incapable of delivering objectively valid determinations of welfare and right that could be susceptible, let alone worthy, of legal enforcement.

Brudner claims that the egalitarian, welfarist paradigm fueling one side of the crisis in tort and contract law advances a system of positive rights promoting the welfare of a realized human capacity for autonomous action and that these rights rest upon three types of entitlements arising from Hegel's account of morality: a right of intention, a right of insight and a right to welfare. [98] Hegel himself characterizes the right of intention in purely moral terms as the entitlement to be held accountable only for those consequences of what one does on purpose that are prefigured in one's motive.[2] This right follows upon, and incorporates, the elementary moral right to be blamed only for the part of one's deed, the action proper, that reflects the agent's purpose. Both the rights of purpose and of intention are complemented by a right of insight that further circumscribes the scope of moral responsibility to those purposeful actions and intended consequences whose own character and circumstances are known by the agent.[3] Since these three rights together oblige all moral subjects to hold one another accountable in function of their own insight into the situation of their action and of their self-selected purposes and motives, they all thereby enjoy the entitlement to pursue ends in which they have a personal interest, thereby realizing a welfare consisting in the fulfillment of that self-selected interest.[4] Hegel emphasizes that each of these rights is wholly formal, since being blameworthy only for what one knowingly does on purpose and causes to result as intended says nothing about what content the chosen purposes, motives, and corresponding welfare will have. All can conceivably conflict with right, which is why morality must next address the de jure question of what purposes, motives, and welfare are in accord with right, a task that conscience undertakes in the morally specific terms that the good to be sought has no objectively given reality and must be determined personally by the moral subject.[5] If, on the contrary, the good had a predetermined objective content, conscience would not be in play. Instead of facing the "disengaged" predicament where moral reflection must be called upon, agents would find themselves obliged to fulfill a commonly recognized preexisting good in which their welfare would be joined to right in the pursuit of preordained roles that reproduce that good. In conditions of moral self-determination, by contrast, agents face the challenge of subjectively determining a good that neither already exists nor has any preestablished content. As a consequence, each conscience confronts the dilemma of either recognizing the differing moral determinations of others at the expense of relinquishing its own moral authority or following the commands of its own conscience at the cost of withdrawing recognition of the authority of the diverging conscience of others. Hegel emphasizes that this quandary has no solutions so long as the moral standpoint is not superseded.[6] What alone removes the endemic controversy of moral reflection are structures of ethical community, wherein individuals exercise institutional freedoms that can only be performed in belonging to an association where the ends members can pursue in so determining themselves are automatically coordinated in a common reproduction of the community whose existence is both presupposition and product of their activity. In the family, spouses enjoy the right to codetermine household property and to have their welfare jointly cared for only by participating in a marital union that exists through nothing but the ongoing fulfillment of conjugal roles. Similarly, in civil society, market participants exercise their economic freedom only by entering into the interdependent network of commodity exchange where the satisfaction of self-selected needs for commodities can only occur by facilitating the complementary satisfaction of another's market needs. Finally, citizens can only exercise their political freedom by performing roles, such as voting or running for office, that depend upon other citizens engaging in the same political participations, without which the institutions of self-government collapse.

Whereas each of these ethical freedoms involves the pursuit of ends whose achievement comprises a welfare in objective conformity with right, such objective conformity is precisely what is lacking in the moral rights of intention, insight, and welfare. Although Brudner wishes the right of intention to liberate rational agents from any forces that do not embody their own goals, [98] this emancipation would undermine every right of property, family, social and political relations insofar as these goals have no predetermined limits that would guarantee their accord with any other form of freedom. Because the right of intention leaves completely open what interests motivate each agent, the right of welfare can only be considered a right to the material and cultural conditions for autonomy [98] if the autonomy in question is recognized to be a purely personal creation, whose realization may well violate the property and ethical rights of others. If the content of welfare was instead defined in terms of the specific requirements of family, civil and political freedom, this license would be overcome.

As by now might be expected, the disanalogy in Brudner's use of Hegel reaches its culmination in the introduction of dialogical community. Brudner explicitly identifies dialogical community with Spirit and appeals to the master-slave relation described in Chapter IV of the *Phenomenology of Spirit* as the paradigm for comprehending Spirit. On this basis, Brudner describes dialogical community in terms of reciprocal recognition, with the twist that the parties to mutual acknowledgment are not individuals but individuals on the one hand and the community on the other. Significantly, the community is referred to as an abstract unitary entity, without any differentiation of family, civil society and state and the distinct goods of each. The common law here allegedly achieves coherent unification by individuals recognizing the welfarist claims of the common good in exchange for recognition by the community of their formal right. Failing that mutual recognition, neither party can uphold its respective principle.

Given how Brudner has identified the two competing elements in common law with abstract right and morality, their integration in dialogical community stands in stark contrast to the unification that Hegel himself offers in the *Philosophy of Right*. Hegel does not and cannot present reciprocal recognition as an encompassing unity for the simple reason that every relation of right, from property and morality through family, civil society and the state, constitutively involves a recognition process in which its specific rights and duties are sustained. Accordingly, the contrast between property right and morality, on the one hand, and the structure into which they are integrated, on the other, cannot be drawn in terms of any distinction between atomistic relations and an association of reciprocal recognition. No structure of right is atomistic and every structure incorporates dialogic association. Significantly, when Hegel introduces the type of freedom that overcomes the complementary limitations of abstract right and morality, he identifies it as ethical community, which he then develops in three successive shapes under the rubric of the family, civil society and the state. Moreover, Hegel situates the administration of law within the ethical community of civil society, implying that "common law" will reflect the very concrete considerations of right in civil society, which include not only upholding person and property, but also maintaining the domestic welfare of spouses and children, as well as the economic welfare of members of the market. The political dimension of law, including legislation and constitutionality, fall within the ethical community of the state, in which property relations undergo a final shaping to cohere with equal political opportunity and the institutions of self-government. At each level, property rights and moral autonomy remain recognized, while subjected to *partial* restraints, for the simple reason that if individuals are deprived of their status as persons and moral subjects they become tantamount to slaves, who are then incapable of exercising any domestic, social or political freedom. These fundamental contours of Hegel's account of right suggest that the reconciliation of "formal right" and the "good" in common law ultimately rests upon the freedoms of ethical community. These integrate ownership and moral accountability with the rights of membership in the family, civil society, and the state through the administration of law in civil society, as governed by legislation and constitutionality. Unless "dialogical community" signifies much more than the mere concept of spirit or reciprocal recognition, it neither fits Hegel's own resolution to the limits of abstract right and morality, nor supplies a sufficiently determinate basis for explaining the complementary elements of the common law. This inadequacy becomes evident when the "crises" afflicting each sphere of the common law are examined in turn.

The starting point of Brudner's immanent critique of the crisis in common law is his analysis of how formal right's realization in property relations is self-contradictory, thereby refuting the formal freedom paradigm as a self-sufficient basis of right. [69] Brudner charts the alleged self-contradiction in property by characterizing property right as first appearing as a solipsistic relation between a person and a thing, in which one affirms one's capacity of choice independently of all connection to others. [58, 59] For the property owner, right allegedly consists in permission to appropriate the world as an embodiment of the choosing self [69], with each particular will remaining free to acquire others' possessions with or without consent and to break contract agreements at will. [70] Brudner claims that property relations end up showing that the person's abstract capacity for choice cannot have the unconditioned worth that ownership affirms because property depends upon the forbearance of others who are incipient violators of ownership rights. [70] This supposedly reveals that the process of embodiment, first regarded as extrinsic to the free will, is essential to it, expanding right into an embodied freedom encompassing positive as well as negative rights, rights that limit an owner's takings of property by the equal right of all persons to the material prerequisites of self-determination. [71, 74] Brudner offers this "self-contradiction" as nothing but the outcome of the application of the standard of mutual respect to property relations. [72]

Each step in this "self-refutation" of formal right has a peculiar ring. To begin with, property right is never reducible to a solipsistic relation of the individual to a thing, expressing the general capacity of choice. In order for ownership to be distinguished from physical possession, a process of mutual respect must enter in by which a person has its will *embodied* in a recognizable way in some factor that other persons recognize as the particular external domain of that will, individuated by its contrast to the recognized embodiments of the wills of other persons, each of whom owns property by having its willing similarly mediated by that of others. Because individuals first determine themselves as individual owners through this interaction, their freedom as persons *always* involves a recognizable embodiment, which cannot have any determinate reality apart from recognition of the property of others. Although persons all retain the capacity of choice enabling them to violate the right of others, as well as dispute the boundaries of their respective domains, they cannot uphold their own ownership without acknowledging that of their counterparts. By itself, the recognition process of ownership makes no reference to the particular interests or motives of individuals; it simply concerns the recognized embodiment of the will of each owner, an embodiment first enabling each individual to give itself that artificial self-determined agency by which owners stand distinguished from slaves. Consequently, in mutually respecting one another *qua* owners, persons do not have regard for how their ownership affects independently given concerns, such as family, social, or political welfare. Yet, because exercising the agency of owner rests upon certain physiological and psychological prerequisites, insuring that *these* are not undermined is a proper and consistent application of the respect of property right. Consequently, the right of necessity, whereby individuals are permitted to infringe on the property of others when their survival requires, completely accords with the formal right of property since it allows the preserving of the entire capacity for ownership to trump a particular infringement of property that can later be compensated. A duty to rescue can similarly be associated with property right to the degree that respect for persons entails action to save another when one is specially situated to give life-saving aid without unreasonably jeopardizing one's own capacity of ownership. So, too, can property right consistently accommodate a nuisance prohibition that limits the use of property when it threatens the property of others. Like the right of necessity, such narrowly framed good Samaritan duty and nuisance restrictions involve no positive concern for anything but safeguarding property relations and their prerequisites.

Accordingly, if the common law is to impose restrictions upon takings of property that go legitimately beyond these consistent applications of property rights, the source will lie not in internal inconsistencies, but in additional entitlements that supervene when property relations are incorporated within further institutions of freedom, such as the family, civil society and the state. Insofar as the empirical reality of the common law applies to individuals immersed in all their concrete relationships, it should come as no surprise that the historically given common law reflects much more than the basic imperatives of formal right.

Although Brudner poses the problem of the common law as if the competing dimensions have an equiprimordial character, the welfarist, egalitarian paradigm has a decidedly posterior position that enables Brudner to puncture effectively any attempts to promote the common good without acknowledgment of formal right. Contrary to all those thinkers, from Locke through Feinberg, who make property originally conditional upon respect for interests, Brudner makes the fundamental point that dictates that corrective justice precede distributive justice: namely, no common good can be the good *of* the individual unless the individual is already recognized to have a domain secured against others' advantage. [78] In other words, property right is basic, because without its realization, nothing meted out to individuals can count as theirs. Consequently, whereas property right can be conceived and realized without incorporating other normative considerations, no other rights and duties can be coherently thought or actualized unless they contain space for the interaction of owners. This does not mean that property right remains unaffected by being encompassed by other structures of freedom; it simply gives expression to the basic truth that unless agents are recognized to be owners of at least their own bodies, they cannot exercise any further rights.

When Brudner turns to contract and tort law to follow out how they, too, are torn by opposing paradigms that point towards their integration in dialogical community, a similar asymmetry prevails.

Brudner claims that each side of contract law self-destructs: when the paradigm of formal right exclusively governs contract, it allegedly subverts property entitlements, which end up requiring equity to fulfill them; when, alternately, equity reigns supreme, domination is reinstated. [99] The latter outcome is readily argued, as one more application of the dependence of distributive justice upon the recognition of personhood. The self-refutation of a formal right contract doctrine is less compelling, since contract can be developed simply as a relation of property transfer, as Hegel, for instance, conceives it as the second theme of property right,[7] for which any discrepancy with the common law represents cause for finding external distributive standards (such as family, social and political welfare could mandate) rather than an internal incoherence in property exchange. Brudner, in effect, acknowledges the asymmetry in arguing that only formal right can be judicially vindicated: if a court were to enforce the right of welfare as the governing principle of contract, it would eliminate contractual right entirely, whereas if legislatures actualize equity through taxation, they uphold ownership by redistributing property originally allocated through contract unrestricted by distributive concerns. [151]

Brudner's case for tort law [153 ff.] is equally uneven. He claims that formal right cannot ground a noninstrumental tort law by itself because the formalist paradigm leads necessarily to governing tort in respect to a common good. [166] Brudner locates this inexorable move in a supposed latent contradiction involved in protecting the embodiments of freedom in a person's body and external property. [175] The contradiction is allegedly twofold: on the one hand, upholding personal sovereignty over body and property conflicts with the affirmation of the person as an absolute end, since it treats the self as absolute only as embodied; on the other hand, the self's need for embodiment contradicts formal right's alleged indifference to the well-being of the person's body and to the quantity and condition of its property. [175] As evidence of these inconsistencies, Brudner cites three fixtures of tort law: 1) the entitlement of victims of trespass to real damages for any resulting harm to body or property, [175] 2) the entitlement of victims of negligence to recover damages unconnected to interference or trespass [176], and 3) the defense of necessity. [177]

Each of these "inconsistencies" with formal right rests upon a confusion in the conception of the person, according to which the agency of property right is conceivable apart from embodiment. This separation may be true of the natural will, whose given capacity of choice is a potential, rather than an actuality, preceding every act of will. The person, however, who achieves self-determination as an owner only possesses that artificial agency by being recognized to have externalized its will first, in its own body, and then, in other external factors. Although the person is always already individuated by natural features it has not chosen, the determinacy that the person *qua* owner gives itself lies nowhere else than in its property. Consequently, when body and property suffer damage, the person has suffered injury with a magnitude that can be measured by the condition of these external factors. Hegel makes this point in connection with the determination of compensation and punishment in cases of non-malicious wrong, fraud and crime.[8] Although Hegel's treatment of non-malicious wrong makes no explicit mention of tort or negligence,[9] the link of person to embodiment equally bears upon the entitlement of real damages for harm to body and property resulting from trespass as well as from negligence. In both of these cases, the person is directly touched by damage to its embodiment and therefore compensation is required. When the damage is a result of negligence, punishment should equally be mandated, since negligence is only distinguishable from cases of non-malicious accidents when the malefactor is held responsible for creating conditions that place the body and property of other persons in unreasonable risk. Real damages for non-malicious accidents represent a case of strict liability, where compensation is called for independently of misfeasance insofar as the accident is caused by the property of an individual, who, as owner, is thereby responsible. By contrast, negligence involves conduct that shows lack of respect for the personhood of others. Therefore, it involves misfeasance to which punishment must respond, in addition to compensation for any harm done. Although Brudner maintains that harm is essential to an action for negligence, [189] endorsing much given practice, the misfeasance in negligence is just as punishable when no harm occurs as are attempted crimes that fail to produce victims. The wrong of negligence therefore does not rest upon a right to affirmative care for a welfare extending beyond the prerogatives of personhood. By jeopardizing a person's health, bodily integrity and possession, negligence strikes at nothing more than what falls within the scope of property right. Admittedly, the scope of negligence can and should be enlarged when the further entitlements of family, social and political community come into play. Nevertheless, contrary to Brudner's claims of inconsistency, [189] the abstract right of property relations can coherently incorporate a more circumscribed negligence principle together with compensation for real damages from trespass and a right of necessity.

Brudner's argument against the absolutization of social insurance is more cogent, although somewhat overstated. If social insurance were entirely to preempt tort law, Brudner maintains, individuals would lose the opportunity to redress wrong *qua* wrong and vindicate the worth of their person against violations. [206] Wrongdoers would no longer be certified as such by being required to compensate for losses they produce and victims would no longer receive an authoritative acknowledgment that their rights have been infringed upon. [206] If, however, the tort involves non-malicious wrong, the social insurance provided to the victim could be said to acknowledge the rights of the victim, especially when the costs and uncertainties of litigation may deprive victims of the prompt compensation they are due. Moreover, so long as the person responsible for compensation under formal right has acted non-maliciously, the payment of compensation eliminates any abiding wrong. Only if negligence or crime were involved would social insurance compensation of victims leave a wrong still standing, namely the malicious wrong residing in the willing against right of the malefactor. That wrong is only erased through punishment that specifically nullifies the misfeasance by constraining the wrongdoer's will through confinement. These considerations do not justify the exclusive hegemony of social insurance, but lend support to Brudner's own endorsement of a dual system combining social insurance and tort law. [205]

Brudner's final foray into penal law reflects the same ambiguities that plague his attempt to unify the common law. Once more, his critique of an absolutization of the welfarist paradigm has more force than his parallel unmasking of the self-contradiction of the formal right paradigm. As Brudner observes, if crimes are reinterpreted as harms, as in Feinberg's identification of wrongs as setbacks to interest, the rational for punishment is undermined, since compensation and prevention of harm preempt any need for retribution or assignment of guilt. [216] Brudner confuses matters by identifying negligence as the standard of fault in the welfarist paradigm [216], neglecting how negligence involves punishable misfeasance. He goes most astray, however, when he turns to the pure agency model of penal law and claims that it cannot differentiate crimes according to seriousness without appealing to welfare considerations. [217] This claim rests upon Brudner's separation of embodiment from personhood and the resulting failure to recognize how harm to the body and property of persons has a magnitude that corresponds to the seriousness of the violation of abstract right. Admittedly, within the context of civil society, crime takes on the additional significance of a violation of social right, whose seriousness is, as Hegel points out,[10] relative to the stability of the social order. Here questions of deterrence properly enter in, but not at the expense of canceling the significance of crime as a malicious violation of the abstract right of the person.

Once more, the resolution of competing strains in the common law rests upon the integration of legality within the institutions of ethical community. Only if the legitimate shape of these formations is examined in detail can the due modification of property right be established. Although Brudner's work repeatedly points to this undertaking, tackling the venture requires leaving behind the abstraction of dialogical community and conceiving law in the context of the ethical associations of family, civil society,[11] and the state. Whether the common law has unity, and more importantly, a unity worthy of adherence, will depend upon the outcome of that enterprise.

*****

The family comprises the first and most elementary institution of ethical community, presupposed and incorporated within both civil society and the state. Determining the family as an ethical association requires, however, that household relations be distinguished from both contract and friendship, as well as emancipated from traditional kinship associations. The following chapter undertakes the conceptual demarcations on which this enterprise rests.

Notes

1 Brudner, Alan, *The Unity of the Common Law: Studies in Hegelian Jurisprudence* (Berkeley: University of California Press, 1995). Numbers in brackets refer to the pagination of this text.
2 Hegel, G.W.F., *Elements of the Philosophy of Right*, ed. by Allen W. Wood, trans. H.B. Nisbet, {Cambridge, UK: Cambridge University Press, 1991}, paragraph 121 ff., p. 149 ff.
3 Hegel, paragraph 117, p. 144.
4 Hegel, paragraphs 121–123, pp. 149–50.
5 Hegel, paragraph 131, p. 158.
6 Hegel, paragraphs, 137, 141, pp. 164–5, 185–6.
7 Hegel, paragraphs 72–80, pp. 104–113.
8 Hegel, paragraphs 90, 91, 96, 101, pp. 119–20, 122–3, 127.
9 Hegel does address tort liability in paragraph 116, so as to contrast the strict liability of owners for damage caused by their property with the more limited responsibility of the moral agent, who, as paragraph 117 explains, can only be blamed for what is done on purpose with knowledge of the circumstances of the action. See Hegel, paragraph 116–7, p. 144.
10 Hegel, paragraph 218, pp. 250–1.
11 I have attempted to fulfill part of this project in *Law in Civil Society* (Lawrence, KS: University Press of Kansas, 1995).

Ten

Friendship, Family and Ethical Community

Hardly any ethical question cuts so near and dear as that of how the family should be ordered. Whereas in time of peace, social and political affairs can hover remotely beyond daily life, controversy over family values directly strikes home, invading the private sphere from which no haven can be found.

Far from being an eternal predicament, this accentuated prominence of family affairs reflects the ongoing upheavals of modernity. With the accelerating destruction of traditions privileging birth right, gender, race and other natural[1] distinctions before self-determination, with the accompanying demarcation of social and political domains,[2] and with the extrication of both civil society and state from kinship relations, the household has won a disengaged existence in which family values can figure as autonomous issues and individuals can enjoy a private sphere that only family matters immediately address.

Nevertheless, as much as the liberation of economic and political power from household bonds has transformed the family into an association with concerns radically its own, the possibility of rationally judging between competing ideals of household life has remained mired in dispute.

Obstacles to an Independent Family Ethics

The road to an independent family ethics seems barred by the intractable resistance of household relations to any ethical reconstruction. Unlike society or state, the family appears to be directly rooted in biological requirements of sexuality, procreation, and upbringing, making household association a sphere of natural necessity from which the freedom constitutive of ethics seems excluded.[3] For if the family comprises an organ of species being, structured in function of sexual difference, reproduction and nurturing, family members can hardly be responsible for the fundamental shape of household community. Life in the family becomes an anthropological fate, whose flourishing may be judged in physiological terms of health and disease or by evolutionary criteria of fitness for survival, but not by ethical standards of right and wrong.[4]

The banishing of the family beyond good and evil seems equally inescapable on the rampant psychological view that the household is a crucible of personality formation within which psyches receive their emotive identity and experience either emotional fulfillment or distress. The psychological privileging of the family in personality development may reflect the disengagement of the household from society and state and the resulting isolation of young children from anyone but immediate family members.[5] Nevertheless, this leaves the family a playing field of psychological mechanisms, in which autonomy figures only as the outcome of proper personality development. Once more, talk of rights and duties fades before issues of health, where now what matters are the therapeutic questions of whether family members adequately promote one another's mental well-being.

Admittedly, the presence of biological and psychological dimensions in family life need not exclude an ethics of the household so long as what distinguish family association are not merely physiological and psychological processes. After all, neither reproductive biology nor personality formation automatically dictate how family members define their interrelations. Where then might the element lie by which the family warrants norms of its own?

Love seems to be the likely candidate, setting household bonds apart as much from biological and psychological compulsions as from the impersonal ties of other normative spheres. Yet how can love, imbued with feeling and subjective attraction, be a pillar of family right and duty? With all its arbitrariness and transience, can love be a basis for distinguishing an ethical household from romantic attachments, where lovers may be subject to the moral imperatives of intimate friendship, but otherwise escape the rights and duties of belonging to a common home? Love, like physiology and psychology, may well be one element in family life, but can it mandate the norms governing control of family property, provision of care, childrearing or any other constitutive family affairs?[6] Whether Platonic, romantic, or simply erotic in character, love can hardly define family association if love relations can occur independently of household membership. Indeed, under the banner of freedom of the heart, love can be the ground for abandoning children to public institutions[7] and ignoring every marriage commitment. Like any other factor that can be differentiated from the family, love cannot serve to determine household norms when it leaves unspecified its own relation to the family.

A similar dilemma undermines any attempt to ground a family ethic in functional terms. Determining principles of household conduct by how they serve independently given aims always leaves unanswerable why families should be retained in the first place, as well as what shape the family ought to have. Since the ends for whose sake the family is to be fashioned are defined apart from family association, their fulfillment is always possible by other means, including alternate household arrangements. Moreover, with family organization subjected to external provisos, the norms of family conduct lose their specificity, merging with every other recipe falling within the general technique of choosing means to achieve the privileged goals to which private life is subordinated.

No solution is provided by turning to the standard alternatives to functional accounts of conduct: namely, the relations of formal right, comprising the sphere of property and contract, and those of moral accountability.

The Incongruence of Family Ethics with Formal Right, Morality, and Politics

The appeal to formal right seems attractive, both because modern marriage has come to rest upon the consent of the marriage partners and because notions of fair contract seem to offer standards for reforming family relations in line with prized aspirations of freedom and equality.[8] The example of Kant highlights the anomalies inherent in taking the first step toward reducing family ethics to formal right. To fit marriage into the mold of contract, Kant is compelled to argue that spouses mutually consent to the exclusive reciprocal use of one another's body, entering a relation in which persons figure as factors of property, subjecting themselves to treatment as things.[9] Although these provisions verge upon a mutual enslavement undermining the autonomy necessary for entering into contract,[10] Kant cannot refrain from postulating them since any more restricted contractual language would reduce marriage to a limited enterprise indistinguishable from a mere exchange of particular services. Kant here runs up against the fundamental stumbling block of any attempt to conceive family ethics in terms of formal right: the juridical independence underlying contractual relations is precisely what the unity of the family submerges by constituting a common private sphere to which one must first belong in order to enjoy the rights and duties differentiating domestic life from relations among outsiders. To the degree that spouses join into a single unified person, interacting without any intervening terms, the conjugal relation cannot be reduced to the formal right of contract,[11] which constitutively concerns interaction through disposition over property, i.e. factors external to persons.[12] Accordingly, when Kant attempts to subsume marriage under contract, he must reduce the relation of spouses to a relation of their external facticity, i.e. their bodies. Given the substantial unity of conjugal ties, the contractual relation turns into a total appropriation of each spouse's body by the other party, undermining the juridical independence of the contracting individuals and thereby subverting the entire marriage contract itself.[13]

Consequently, when feminists like Susan Moller Okin[14] prescribe how relations within and without the family should be reformed to enable spouses to marry in genuine conformity with formal right, they appeal to norms of contractual liberty that cannot lay hold of what is distinctive in family ties.

Any attempt to derive family norms from liberal notions of distributive justice falls into the same dilemma. Distributive justice might seem to have ample scope in family relations, given the incidence of dowries, gifts to kin, inheritance, alimony, and other provisions of mutual aid.[15] Nonetheless, because the procedural justice of social contract thought always privileges the individual liberty and separate self-selected interests of persons,[16] liberal theory cannot coherently acknowledge how membership in a family could confer rights and duties that are irreducible to calculations of self-interest, property entitlements, or applications of a categorical imperative.[17] Indeed, within the family, the very circumstances of justice, as construed by liberal theory, have been supplanted.[18] As Hume and Rousseau already recognized, with, "all distinction of property ... in a great measure lost and confounded"[19] in the trust, affection and unity of ends distinguishing family association, the independence and opposition of individual interests is already preempted without the intervention of social contract and civil government.[20] However, although liberal thinkers like Hume and Rousseau may conclude that the family makes standards of civil justice irrelevant by already realizing the harmony of aims, they ignore how the family involves interests and goods qualitatively distinct from what can enter into the atomistic considerations of any utilitarian calculus or social contract.

Admittedly, family members remain discrete persons with distinct goals and ambitions that may well conflict. Yet, this does not reduce their rights and duties to one another to those that apply between members of society who do not belong to the same family.[21] The love and intimacy of family relations are not liable to redistribution on the same terms as property or any other "primary goods" precisely because family ties rest upon a unity of obligatory favoritism drawing a sharp divide between family members and outsiders.[22] For this reason, it makes little sense to adopt the program suggested by Okin of conceiving the just family by determining what persons under a veil of ignorance would agree to concerning domestic organization.[23] The embrace of liberty and rational self-interest underlying procedural justice simply provides no compelling rationale for instituting families in the first place, especially when family ties introduce parochial attachments undermining any patterned distribution to individuals.[24] Therefore, it is hardly surprising that liberal theorists like John Rawls[25] and James Fishkin[26] suspect that the family poses an insurmountable impediment to distributive justice, engendering an insoluble trilemma where fair competition and equal opportunity cannot be upheld if the autonomy of the family is simultaneously respected.[27]

To the degree that family members have special ties to one another, morality is no more amenable a resource than contract for providing ethical sanction to distinctively domestic relations. Because moral subjects hold one another responsible exclusively for those aspects of their deeds prefigured by their personal choice of conscientious purpose and intention, the rights and duties of moral accountability extend to all agents, irrespective of their membership in any existing associations. From an exclusively moral point of view, a family ethic is just as unimaginable as a political ethic that differentiates between what rights and duties citizens share and what rights and duties apply between citizens and non-citizens.

Yet if the particular boundary of the family makes its ethical standing problematic for those who model normativity after formal right and morality, a parallel dilemma faces those who conceive of a good life as coextensive with the political life of the citizen in the best state. For, as Plato recognized, if conduct has its exclusive worth in contributing to the common good enshrined in the body politic, then families are dispensable evils in virtue of the divisive, parochial loyalties they foster.

Relativism, Nihilism and the Multi-cultural Challenge to Family Ethics

Whether privileging formal right, morality, "economic rationality" or politics, prevailing ethical theories leave no room for conferring any independent worth upon the family. Either ethical conduct is incongruent with the particular, substantial ties of domestic life or it enters into family affairs only in those recesses that involve norms given independently of household bonds.

This lingering theoretical abstention might seem to fly against modern developments that have disengaged the family from economic and political affairs by replacing household autarchy and hereditary occupations with market relations operating independently of kinship, by instituting social welfare institutions diminishing the role of household care and inheritance as bulwarks of livelihood, and by erecting states in which political power is shielded from birthright and nepotism. Yet these same modern developments have been accompanied by other transformations of family organization that have called into question what if any norms pertain to relations between spouses and between parents and children. With the increasing prevalence of no-fault divorce on demand, the spread of prenuptial agreements limiting common household property, and the growing legal recognition of unwed couples of any gender and sexual orientation, the very significance of marriage has been cast in doubt. Similarly, the resulting increase in single parent households, in public day care of children of all ages, in premarital sexual activity, and in the intervention of courts and welfare bureaucracies on upbringing, have all raised questions about what parent-child relations should be. And the very rationale for the family has become clouded with the growing proliferation of single persons, be they adult children living on their own, women postponing or avoiding marriage and childrearing to pursue educational and career opportunities, self-sufficient widowed "senior citizens", or any others for whom available housing, economic independence, and social welfare guarantees make life without kinship a secure and untarnished option.[28]

If the natural and psychological dimensions of the family resist ethical incorporation, if the most popular ethical theories find no place for family values, and if the boundaries of the modern family themselves appear in flux, can a family ethic be established at all? Are norms of family organization merely cultural conventions, defying rational justification and undeserving of any hegemonic ranking and regulation?

The multicultural bandwagon might appear to support an unconstrained pluralism in family practices, especially if toleration of cultural diversity gets tied to the postmodern dogma that reason and conduct are always hostage to historically given practices, so that any attempt at justifying knowledge claims and ethical norms must always appeal to privileged concepts and vocabularies reflecting interests, power formations, or other pragmatic concerns. Why not accept the coexistence of alternate family regimes, each representing the values of a particular culture or subculture?

Yet can multiculturalism signify a blanket toleration of all cultural practices, including all family forms? The post-modern argument for multicultural pluralism tends to vitiate itself, both because too few cultures practice toleration, making implementation of multiculturalism a contradictory enterprise, and because postmodernism relies upon reason to discredit rational justification while making its own global claims about the foundation-ridden character of discourse and conduct.

If multiculturalism is to have any coherent justification or implementation, it must instead signify a modern, rather than postmodern, embrace of the principle that the realization of self-determination requires that property relations, moral accountability, household affairs, society and the state all be liberated of bondage to cultural differences, so that ethnic, linguistic and religious identity have no more role in restricting the opportunities of individuals in their family, social or political involvements than birth, race, gender or sexual orientation.

Upholding this modern agenda requires first, justifying the exclusive normativity of self-determination, and then, second, establishing what the family must be to comprise an institution of freedom, congruent with the other spheres of right. The travails of liberal theory exhibit the difficulties of achieving either task. Justifying the legitimacy of freedom remains a futile endeavor as long as appeal is made to prior factors, such as the internal collapse of teleological ethics, facts of nature or self-interest. Any such attempt runs up against the dilemma undermining foundational argument – by distinguishing what confers legitimacy from what possesses legitimacy, foundationalism insures that the factor begetting validity can never exhibit what counts as validity.

Freedom, Ethical Community, and Family Ethics

Freedom is immune to this problem because self-determination comprises something that cannot owe its character to anything but itself. Escaping all juridical dependence upon antecedent givens, self-determination can command an unparalleled legitimacy residing in the two complementary sides of its freedom from foundations: on the one hand, the negative liberation from the given, whereby what is specific to freedom is undetermined by any prior factors, and, on the other hand, the positive autonomy wherein self-determination gives itself its own character precisely by being what it determines itself to be. Unlike any foundationally justified practice, self-determination is therefore, normatively speaking, presuppositionless and self-grounding, eliminating the gap between justifying factor and justified conduct.

Accordingly, if self-determination is to be properly conceived, it cannot be thought in liberal fashion as a given principle determining something other than itself, where the form of autonomy is given either by nature as the capacity of choice underlying all conduct or by abstraction as a privileged choice procedure from which legitimate institutions are derived. In both cases, freedom's own character would be defined antecedently to all acts of will, whereas whatever conduct its principle would sanction would equally owe its determination to a privileged factor. Hence, neither freedom as principle nor its derivative domain of conduct would be self-determined. In order for self-determination to give itself its own character, the very agency of freedom must issue from its willing just as much as the content of what it wills. Then, the form and content of autonomous agency will be products of free willing, engendering an artificial agency enabling conduct to figure as the self-determination of self-determined agency.

For conduct to succeed in defining its own agency and thereby achieve genuine self-determination, individuals must act in respect to the volition of one another, engaging in conventional practices within which they can first exercise types of freedom comprising those enabling practices themselves. By willing one's relation to others so as to allow them to will their relation to other agents as well, one can individuate one's own agency through one's willing, as one among other self-determined individuals. Such artificial agency is at work in every exercise of rights, since rights, in contrast to privileges, can only be enjoyed by participating in a framework within which all other participants in principle respect each others' entitled freedoms by willing in conformity with one another. In such a context of interrelated conduct, individuals are able to determine themselves as holders of the rights involved, e.g. as property owners, moral subjects, market agents, legal subject, or citizens, by observing the rights and duties whose performance comprises the enabling context in question.

For when individuals interact in respect of one another's willing, they are able to occupy a non-natural, conventional role exercising an intersubjectively recognized freedom defined by the reciprocally connected volitions that build their conventional practice and individuate them as participants. Far from being a natural given or a function of the self, self-determination must thus consist in modes of interaction where each participant's exercise of freedom involves respecting the like exercise of freedom by others, who equally determine themselves in recognition of the congruent autonomy of their peers. In the ensuing web of right and duty, freedom will not stand apart from institutions of justice as their prior principle but rather comprise a system of institutions of freedom whose artificial agencies will inhabit different spheres of right, each involving its own distinctive entitled prerogatives, yet integrating with one another into a single self-ordered whole. The very normativity of freedom requires such an integration, for otherwise different freedoms would limit one another for the sake of some ground external to self-determination and the spheres of right would find themselves connected by an order imposed upon them all, rendering them dependent rather than free.

If the family can be structured as one such integrated sphere of right, wherein individuals exercise rights and duties as autonomous family members that are nowhere else available, the possibility of a discrete family ethic will be secured. This ethic will consist in realizing those entitled freedoms by which spouses, parents and children engage in household roles affording them a mode of self-determination otherwise unattainable. Only then will a domain of family values be established free of the pitfalls of foundationalism that have subverted other strategies for upholding the normativity of household relations.

Yet, how must ties between spouses, parents and children, siblings or other kin be configured for the family to comprise an institution of freedom with its own non-arbitrary rights and duties? If the normativity of the family is to be secured by conceiving a distinct sphere of household freedom, the frameworks of property and moral relations must be left behind, together with the ethical theories that have privileged them. Instead a structure of freedom must be found that can accommodate a mode of self-determination in which agents determine themselves as members of a joint private domain with its own characteristic good and corresponding rights and duties. The prime resource for providing such a framework is ethical community, which has increasingly become a focus of attention more for accommodating social and political freedom than for encompassing family self-determination. However, ethical community cannot be immediately appropriated for developing the normativity of the household. The deformations of the communitarian turn to ethical community must first be acknowledged and overcome. The communitarian turn may well proceed from due recognition of how teleological ethics stamps conduct with a particularity lacking universality, whereas procedural ethics leaves conduct governed by a universal law devoid of any non-arbitrary particular filling. However, the communitarian formulation of ethical community succumbs to these very problems by affirming 1) that ethical community is the exclusive structure of normative conduct and 2) that the concept of ethical community does not itself mandate the particular content of the roles, ends and association that it may take.

Seeking to overcome the limits of appealing to "the given" or to the form of individual willing as sources of normativity, communitarians have turned to ethical community as an historically contingent structure of shared practices with the same arbitrary content and the same exclusive norm-sustaining role as the language games to which Wittgensteinians turn as the locus of meaning and truth.

Thereby, on the one hand, communitarians absolutize ethical community, rendering the domain of ethics coextensive with the roles animating ethical association, thereby excluding any morality without community as well as any property entitlements that might be given independently of the distributive prerogatives enjoyed through membership in such putative ethical institutions as the family, the market and the state.

Complementing this absolutization is a commitment to the formality of the concept of ethical community. At the same time that ethical community is advanced as the exclusive framework of normative conduct, the idea of that framework is held to leave undetermined what particular institutional configurations and correlative roles and aims it may involve. Since these details are not contained in the concept of ethical community, they are beyond a priori prescription, falling to the grasp of an historical understanding that describes the contingent conventions of history that alone provide content to the otherwise empty idea of ethical community.

This formality is of crucial significance for it directly resurrects the very same problems that the appeal to ethical community is designed to overcome. Just as teleological ethics can not demonstrate the universality of its particular candidate for the good, so communitarianism here renders the content enshrined in ethical community an arbitrary and transient phantom of history. And just as procedural ethics cannot tie the universal form of willing that it privileged to any particular ends without falling back upon teleological appeals, so communitarianism has ended up leaving the universal idea of ethical community powerless to specify any particular institutions to concretize its unity. In effect, the communitarian turn offers little more than an injunction to occupy one's station, an injunction that might satisfy an Eichmann, but otherwise reflects a formality that rivals that of any categorical imperative.

If the concept of ethical community is to provide a solution to the dilemmas plaguing contemporary ethics and serve as a guide to reconstructing modern institutions including the family, the two defining dogmas of the communitarian turn must be supplanted. On the one hand, ethical community can no longer be conceived as a formal framework, whose content is relegated to historical accident. Instead, ethical community must obtain a universally valid content, bringing necessary concreteness to its concept. As the experience of teleological and procedural ethics testifies, such normative content cannot be supplied by appeal to privileged givens masquerading as highest goods or by recourse to privileged forms of willing posing as archimedean procedures of ethical construction. In both cases, factors are immediately advanced as sources of normativity that cannot be legitimated in accord with the very standard of legitimacy that they impose. Whether the legitimating standard be a privileged given content or a privileged determiner of sanctioned ends and institutions, the abiding distinction between legitimated conduct and legitimating factor, a distinction generic to foundational arguments, introduces this debilitating incoherence.

What alone eliminates the distinction and the foundational dilemma it reflects is the identification of normativity with the reality of self-determination. Unlike any other shape of conduct, self-determination owes its measure to nothing but itself, overcoming the gap between legitimating factor and legitimated conduct. In this respect, self-determination alone escapes dependence upon extrinsic foundations and enjoys the self-grounding, presuppositionless character that fundamental normativity requires.[29] Accordingly, if ethical community is to escape the relativity of an externally conditioned content and take its place as a genuine ethical structure, it must be reconceived as an association of freedom, whose agencies, ends and institutions are defined by nothing other than their role as constituents in the reality of self-determination.

However, the rehabilitation of ethical community cannot simply eliminate the formalist assumptions of communitarianism and supplant natural determination with the artificial freedoms of ethical association. The second communitarian dogma absolutizing ethical association as the exclusive mode of normativity must also be overturned. For, far from being coextensive with the reality of freedom, ethical community cannot coherently comprise a structure of self-determination at all unless two other modes of freedom are admitted whose autonomous agencies are not defined by membership in an existing community. These two types of freedom without community are the self-determinations constitutive of property ownership and moral accountability. Ethical community cannot enjoy normativity without recognition of their independent being because both property relations and morality play constitutive roles as necessary preconditions for exercising any freedom as member of an existing association.[30] This is most apparent with respect to the right of persons to be respected as an owner. Exercising this right always involves being recognized as inalienable proprietor of one's own facticity, since otherwise one is liable to be owned by another will like any other rightless factor. Consequently, whenever respect for the right of persons is not incorporated in existing conventions, individuals are in the position of slaves, for whom no act is recognized as juridically their own. In such a situation, the individual can no more be held morally accountable than occupy any institutional role, such as family member, economic agent, legal subject or citizen, to which right and duty apply. The recognition of moral accountability is equally indispensable. If individuals are not entitled to be held responsible for acting with the right purposes and intentions and an understanding of what is good, how can they fulfill any roles as members of a community whose bonds consist in modes of freedom?

Consequently, if the family is to be reconceived as an institution of freedom, its ethical community must incorporate the personhood and moral autonomy of its members. Yet before the family can be systematically distinguished from social and political community, it is helpful to set in focus the household's irreducibility to property and moral interaction by distinguishing friendship from family association.

Friendship Morals Versus Family Ethics

With family norms no more prescribed by the biology of sexual difference and reproduction than by the psychology of childhood development, the family has the same natural and psychological enabling conditions as any other structure of freedom. Conversely, as current convention increasingly brings home, sexual intimacy, reproduction and childrearing can occur just as well among informally attached friends as among formally recognized spouses. This complementary situation brings to the fore the question of how family relations can be differentiated from friendship, for no longer can a distinction be made by appealing to the naturally determined factors of sexual difference, reproduction and nurturing that tradition relied upon. In such an atmosphere it is no wonder that modern law has grown uncertain over whether to extend the same status to friends and their charges as to spouses, parents and children.[31] Nor should it be any surprise that defenders of gay rights can refrain from militating for same-sex marriage, as if matrimony was an expendable freedom and all that really counted was equal treatment in society and state.

Adding to the confusion is the peculiar predicament of the treatment of friendship in the history of ethical theory. Whereas modern theorists from Hobbes to Hegel have let friendship virtually disappear as a distinct topic of ethics, the standardbearers of ancient thought, Plato and Aristotle, have tended to elevate friendship above family relations in the order of the good life.

Their privileging of friendship reflects a ranking resulting from family practices prevailing throughout much of history up until modern times. Friendship has historically been favored on two grounds to which the family could little lay claim until recently. On the one hand, friendship has been recognized to be the field for true romance, where personal bonds are those of freely entered love and sexual adventure. On the other hand, friendship has been acknowledged to forge ties of true equality and reciprocity.

By contrast, the family has gotten assigned a secondary status for just those natural determinations that have marked its traditional organization. On the one hand, the family has figured as a sphere of natural necessity, devoted to procreation and to fulfilling daily recurring subsistence needs. On the other hand, the family has operated as a sphere of naturally determined hierarchy. The relation of spouses has been defined by an unequal gender differentiation, lacking reciprocity, freedom and equality and thereby lacking true romance and love, as well as true respect and self-respect. The parent-child relation, for its part, involves a natural inequality, readily viewed as a technical management, where, as Kant would put it, right applies to persons treated as things.

Against this historical backdrop, the theoretical privileging of friendship over family has an apt resonance, that easily conceals the contingency of the fit. For just this reason, the classical conception of friendship involves its own deformations, whose exposure paves the way for properly distinguishing friendship and family, as well as for comprehending why modern ethicists have tended to give no special norms to ties among friends.

Both Plato and Aristotle assign friendship the role of supplying the surrogate ethical community that a naturally defined, hierarchical family cannot provide. Friendship gets characterized as a freely entered unity, whose participants live their lives in common, benefitting from the joint activity in which what each friend receives is the same good that the other obtains, a good that can be had nowhere else but in the fraternity they share. When this idea of friendship is thought through, it entails withdrawing its participants from any life in a naturally defined household and ends up obliterating any distinction between friendship and a free family, whose members marry by choice and share an ethical love involving reciprocity, self-determination and equality. For if friends are literally to live a life in common, they cannot equally dwell within separate private domains, sharing an *oikos* with wife, slaves and children.[32] Instead, their life in common becomes a life of virtual spouses, provided the natural functions of sexual difference, reproduction, and nurturing of offspring have no constitutive hold on family affairs.

In these respects, Aristotle's description of friendship closely approximates the normative ideal of the family, reconstructed as an ethical association of freedom, whereas his account of the *oikos* straps the household with bonds of natural difference and domination that undermine its ethical community. Aristotle gives expression to these complementary construals by affirming both the identity and difference between family and friendship. On the one hand, friendship and family are alike in that friends have no need of justice,[33] enjoying as they do the same unity of ends characterizing household ties. On the other hand, however, the friendship of family members should be set apart from the friendship of companions because of the inequality marking all but one family relation.[34] Paternal friendship is a tyrannical relation, where a father treats children as slaves,[35] whereas the friendship of husband and wife is aristocratic, reflecting a difference in worth that may verge on oligarchy when the difference gets tied to control of wealth.[36] Yet, under tyranny, Aristotle admits, friendship is hardly possible given the absence of anything in common, while under aristocracy, friendship is less feasible than under democracy, where people are equal and have the most in common.[37] Accordingly, the only family relationship that bears resemblance to complete friendship is the community of siblings, for they are equal except with respect to age.[38] Hence fraternity and friendship deserve the link that language has confirmed, but only in recognition of the divide between true friendship and all the other dimensions of traditional family community.

The reason why friendship is given so much more attention in ancient thought lies in this contrast. For complementary reasons, Plato and Aristotle construe the family in ways that deprive it of proper normativity. For Plato, the family is a sphere of nepotism, whose kin partiality stands implacably opposed to the unity of the polis, where the good of the individual must be at one with the good of the community. Hence, the natural functions traditionally rooted in the family must be withdrawn from the particular home and Plato's abolition of any family for the Guardians confirms how the ends of procreation and subsistence can be achieved through other means.[39] For Aristotle, the family is a sphere of natural domination that may be retained as a subordinate equipment for virtue so long as domestic hierarchy frees the head of the household from attending to necessities of subsistence and upbringing, enabling him to participate in the good life of the citizen.[40]

Yet, by presenting friendship as a superior counterpart to family affairs, Plato and Aristotle are equally at a loss to retain any special standing for friends among the institutions of ethical community. Friendship poses a dilemma for Plato, given his own objections to family parochialism. If friends are partial to one another, but indifferent to the good of strangers, fraternity will collide with concern for the common good and Plato would have to hound friendship out of the polis together with family and poetry. However, if friendship can be viewed simply as living a life together in pursuit of the joint good that its shared life represents, friendship can be redeemed by becoming a universal fellowship, relating all citizens in bonds of fraternity. In that case, however, friendship loses its own specificity, for the virtues of the citizen are now indistinguishable from the virtues of friends. Aristotle, on the other hand, may maintain the difference between friendship and politics by tolerating the particular scope of fellowship among friends. However, because he conceives friendship as a form of joint living, where friends take pleasure in each other for freely sharing in activity together and taking pleasure in the same things, Aristotle can not distinguish between friendship and the form of family he ignores, a free family, whose members choose to become spouses out of love and devote themselves to a life in common embodying reciprocity and equality. Whereas Plato, to be consistent, would have to allow friendship to turn into the universal solidarity of politics, Aristotle ends up reducing friendship to what family affairs would become if freedom and equality replaced natural necessity and gender hierarchy. Either way, friendship retains no independent worth of its own. Instead, relations between friends that do not substitute for family membership and citizenship become no different in ethical character from relations among strangers.

If friendship is to be irreducible to family, social, or political community, it will have to seek its guidelines in formal right or morality, the only alternative modes of self-determination. However, neither property nor moral interaction gives friends any special entitlements or duties to one another. The norms of property relations and moral accountability pertain no differently to friends than to strangers. Neither property entitlements and contractual obligations nor moral duties are contingent upon shared interests and experiences or mutual affection and attraction. Indeed, as Kant emphasizes, the morality of conduct is most evident when one interacts with those one dislikes. Accordingly, it can come as no surprise that friendship receives scant attention among utilitarian and liberal theorists, who model normativity on formal right and morality, to the neglect of ethical community.

Kant might appear to be the big exception to the rule insofar as he maintains that persons have a duty of friendship.[41] However, Kant grounds this duty in the adoption of ideal dispositions making prospective friends worthy of happiness, where these dispositions consist in sharing sympathetically in the well-being of one another by joining in exercising a morally good will. This disposition thereby becomes indistinguishable from the general duty of benevolence that Kant mandates for all persons in their relations with strangers and friends alike.[42]

Friendship deserves no more place of its own in ethics if, like Hegel, one acknowledges formal right and morality, without ignoring ethical community. For, granted that property, morality, family, society and state exhaust the different spheres of right, friendship still engenders no obligations or entitlements specific to itself.

Just as property and moral rights are not conditioned by the friendships of persons and moral subjects, so family relations involve their own entitlements and obligations irrespective of the friendships of their members. Spouses owe each other the same conjugal care and respect whatever other friendships they may have, whereas parents and children retain the same juridical relation whatever their friendships may be. If friendship impinges upon family responsibilities, it thereby wins no juridical place in the home any more than a robber acquires any ownership of what he or she steals.

The freedoms of civil society are equally indifferent to friendship. That commerce involves friends is wholly incidental to its economic character and to individuals' exercise of equal economic opportunity. Similarly, honoring the rights of individuals to equal treatment under the law precludes privileging friends at any stage of the legal process. Amity has no allotted place even in social interest groups, where individuals unite in pursuit of shared ends, for the joint advocacy of such groups is defined by factors specific to civil society (e.g. class interest, consumer rights, environmental concerns, etc.) that are independent of the particular bonds of friendship. By the same token, public welfare institutions care for the economic well-being of individuals without concern for who their friends may be.

Camaraderie is equally incidental to the political arena. Holding political office is based not on amity, but on election by one's fellow citizens, with strangers playing no different a role than friends. Political activity is partisan on the basis of shared political programs rather than on fellowship. If fraternity is to figure as a virtue of a just republic, it must embody the impersonal concern for upholding the rights of all citizens in a constitutional democracy, with utter indifference to the favoritism friends show one another.

In short, if relations among friends have any normative standing, it will not be in virtue of their amity, but in respect to their rights as owners, moral subjects, family members, social agents and citizens. Because none of these rights are qualified by the presence, absence, or degree of friendship among the parties involved, friendship warrants no special treatment in the philosophy of right other than as a contrast term from which genuinely distinctive normative relations can be distinguished.

The differentiation of friendship from family ties serves just this purpose of putting into focus what sets household relations apart from informal fellowship. Friendship distinguishes itself from family membership at every stage of its development. Individuals become friends as a result of the experience of particular mutual decisions to fraternize, without any single agreement that formally ratifies the friendship for the parties involved or for others. Hence, the beginning of friendship has an informal, indeterminate character that only reveals itself retrospectively.

Once established, friendship unfolds through the pursuit of joint activities, but without entailing a common home in which property gets merged and mutual care becomes a strict responsibility. Friends may hold property in common, but that property will be a particular part of their respective holdings, rather than something inclusive of the private domain of each friend. Similarly, friends do not have any claim upon the present and future earnings of one another nor any claim upon their respective properties apart from particular contractual agreements. Moreover, activity on behalf of a friend is a matter of moral benevolence, rather than a juridical duty. Friends are not obliged to be responsible for nor entitled to participate in raising each other's children. Nor do friends who are children have any claim upon or any obligation to obey each other's parents.

By contrast, the dissolution of friendship might appear to closely parallel the dissolution of family ties. Like marriage, friendship may terminate due to the natural contingency of death or the psychological contingency of a change in feelings of attraction and aversion. However, the dissolution of friendship does not raise the problem of dividing a common home since friends already retain separate private domains. Similarly, the end of friendship does not pose questions of alimony.[43] Friends do not belong to a substantial unity for whose upkeep they remain responsible. Ex-friends are not responsible for supporting one another or for supporting the children of ex-friends for whom they are not guardians. Moreover, since friendship imposes no juridical duties requiring legal promulgation and enforcement, its dissolution no more involves a publicly recognized formality than does its inauguration. Indeed, whether friendship is recognized by others is a matter of indifference from start to finish. So long as friends do not infringe upon one another's property, moral, family, social and political rights, their friendship remains a purely private affair, whose integrity includes the prerogative of unilateral termination at any time.[44]

Against these phases of friendship, the outline of family community stands out in sharp relief. Whereas one cannot be placed into a friendship through the acts of others, one commonly[45] first belongs to a family through birth or adoption to individuals who have incurred the publicly recognized obligation of parenthood, an obligation resting not on biological relations of reproduction, but on the acknowledged ethical commitment to raise a child. Alternately, when one does enter a family voluntarily, membership proceeds from a publicly recognized formality, the marriage, with binding consequences for the spouses. Moreover, whereas friendship is entered simply on the basis of mutual willingness, prospective spouses must fulfill formal requirements that today generally involve age, prohibitions of incest and bigamy, and, most questionably, gender.

Once married, spouses distinguish themselves from friends by consolidating their property into a common ownership subservient to the end of maintaining the family well-being, by being obliged to care for the welfare of one another and, when children are involved, by being obliged to insure their upbringing to autonomous maturity. In fulfilling these duties, spouses are entitled to jointly manage the household property and to take joint responsibility for caring for their children. Whereas friends are never compelled to sustain their relation, the entitlements and obligations of family life are publicly enforced.

Finally, when marriage breaks down, a publicly recognized formality, divorce, is necessary to signal the termination of the juridical relation between spouses. Whereas the breakup of friendship remains a private matter, with no publicly enforced consequences beyond what strangers owe one another, divorce entails distributing family property and future income in function of the entitlements of spouse and children.

All of these differences between friendship and family suggest that household association cannot be reduced to the moral ties of friends, where commitment remains subjective and fellowship imposes no rights and duties that individuals do not already possess as owners, moral agents, and members of society and state. By contrast, the ways in which domestic ties diverge from friendship from start to finish indicate that individuals acquire new entitlements and obligations by belonging to their family, for whose common welfare they serve as beneficiary and benefactor alike. As the mores of friendship testify, contract institutes nothing but particular disposals of pre-existing property entitlements just as moral accountability projects the same perennial norms onto each and every relation. By comparison, family ties engender normative commitments that begin and end with family membership, suggesting the basic logic of ethical community, where agents exercise rights and observe duties reproducing the association within which their self-determining role can alone subsist.

Significantly, these differences are not defined by any natural or psychological factors, but only in terms of distinct modes of interaction, involving different rights and duties. For this very reason, the acknowledged discrepancies in friendship and family ties are commonplaces of modern practice, even as the traditional restrictions upon married life and premarital relations go by the wayside. Nonetheless, the familiarity of the divide only delivers an ethical message if it can be supplemented by an authoritative drawing of boundaries, delimiting the structures of freedom. To this end, the abiding differences between friendship and family must be corroborated in the rational reconstruction of the household as an institution of ethical community.

*****

The delineation of family freedom has been notoriously absent in the constitutional guarantees of the most venerable political democracies. The following chapter undertakes the examination of the specific ethical community in which family rights consist, an examination determining the agenda for the revolution in family law still awaiting completion.

Notes

1 These factors are "natural" in the sense of being given, rather than being determined by the will of the individuals to whom they apply. Although their privileging is a matter of convention, they themselves are not artificial products of conduct.

2 The rise of demarcated social and political domains has occurred by overthrowing the feudal relations exemplified in estates where individuals' role in society is tied to political privileges, and where both are tied to birthright. Admittedly, such feudal relations could be eliminated without separating civil society and the state by means of a totalitarian solution where government directly administers the economy.

3 In this vein Aristotle claims that the family is an association rooted in nature from which all others follow (see Aristotle, *Politics*, Book I), Rousseau maintains that the family is the earliest and only natural society (see Rousseau, *On Social Contract*, Chapter II, 8) and Locke proclaims conjugal relations the first society, from which relations of parents and children and then master and servant follow (Locke, *Second Treatise on Government*, Chapter VII, 44).

4 Alan Bloom is one who argues that justice does not apply to the family since nature imposes a hierarchy and gendered division of labor upon household organization. See Okin, Susan Moller, *Justice, Gender, and the Family* (New York: Basic Books, 1989), 26, 36–37, for a critical discussion of Bloom's position.

5 Mark Poster argues analogously that Freudian psychoanalytic theory provides an accurate portrayal of the dynamics of the bourgeois family, but not a conception applicable to other historical family formations where a nuclear patriarchal household is not encountered. See Poster, Mark, *Critical Theory of the Family* (New York: Continuum, 1988). Similarly, Okin points out that psychoanalytic and other psychological theories of personality development have taken gendered parenting as the locus for personality

development and have therein examined how gender is reproduced. See Okin, *Justice, Gender, and the Family*, 131. Freeing psychological theory from such assumptions about gendered parenting would still, however, retain the assumption of the disengaged nuclear family.

6 Hegel raises these issues in his introductory critique of past theories of the family. See Hegel, G.W.F., *Vorlesungen Über Rechtsphilosophie*, Edition Ilting (Stuttgart-Bad Cannstatt: Frommann-Holzboog, 1974), IV:425–429.

7 Walzer, Michael, *Spheres of Justice: A Defense of Pluralism and Equality* (New York: Basic Books, 1983), 238.

8 Conceiving the family in contractual terms might seem to meet an insurmountable obstacle in how children have no choice about into which family they are born or adopted. Yet, liberal theorists since Locke have set aside this apparent difficulty by arguing that children do not yet have an independent capacity of choice and that the purpose of upbringing is to provide them with the conditions for becoming an autonomous individual, capable of owning property and entering into contract.

9 Kant, Immanuel, *The Metaphysics of Morals*, trans. by Mary Gregor (New York: Cambridge University Press, 1991), sections 22–23, AK 276–277.

10 Walzer follows in Kant's footsteps by incoherently claiming that the family is just like the market, except that in domestic affairs the commodities own themselves, as shown in the "gift of self" and the "voluntary exchange of selves" supposedly characterizing parental care and marriage. See Walzer, *Spheres of Justice*, 238.

11 The hedges of prenuptial agreements and the wrangling over property in divorce proceedings do not contravene this submersion of contractual independence in conjugal union. For if formal right enters into household relations at all, it is only in regard to the prospective or ensuing dissolution of the family through divorce or death.

12 Derrida makes this point in his commentary on Hegel's account of marriage in Jacques Derrida, *Glas*, trans. by John P. Leavey, Jr. and Richard Rand (Lincoln, Nebraska: University of Nebraska Press, 1986), 195.

13 Hegel, Georg Wilhelm Friedrich, *Vorlesungen Über Rechtsphilosophie*, edition and commentary by Karl-Heinz Ilting (Stuttgart-Bad Cannstatt:Frommann-Holzboog, 1974), IV 422.

14 Okin, Susan Moller, *Justice, Gender, and the Family*, 121.

15 Walzer, *Spheres of Justice*, 227–228.

16 Social contract theory does this by starting with individuals in a condition antecedent to the enactment of normatively valid institutions, as if the predicament of choice available to them in that "natural" context should somehow mandate how to construct the type of community they should inhabit. Needless to say, this strategy completely abstracts from the framework of ethical community, in which the norms of valid conduct are predicated upon membership in an existing community *already* embodying those norms.

17 Significantly, when Rawls grudgingly identifies the participants of the original position as heads of families, his rationale is that their family membership allows for consideration of the rights of future generations, which will allegedly ground his just savings principle. (See John Rawls, *A Theory of Justice* {Cambridge, MA: Harvard University Press, 1971}, 128–129.) Attending to the rights of future generations, however, is by no means identical with parents' concerns for their *own* children. Moreover, the production and rearing of children need not depend upon the family, as schemes for collective rearing since Plato reflect, nor need the family have children in order to involve the special rights and duties that correspond to what distinguishes domestic association from other forms of community. When Rawls elsewhere brings the family into his theory of justice in any positive way, it is only as a primary school of moral tutelage, that is, as an instrument for achieving independently determined normative aims, an instrument whose own structure is taken for granted. (Ibid., 463, 467 ff.)

18 Michael J. Sandel, *Liberalism and the Limits of Justice* (New York: Cambridge University Press, 1982), 33–34.
19 Cited by Okin in Okin, Susan Moller, *Justice, Gender, and the Family*, 27.
20 Okin observes that the overcoming of distributive justice in the family historically originated in the circumstance that women relinquished their legal personality upon marriage, with all rights to control household property and income passing to the husband. (Okin, *Justice, Gender, and the Family*, 30.) However, with the eventual acknowledgment of both spouses' right to codetermine household affairs, the absence of distributive justice takes on a different character. Moreover, that household property, income, career qualifications and other socially relevant "goods" are still unevenly distributed between husband and wife (as Okin powerfully documents, see Okin, *Justice, Gender, and the Family*, 31ff.) does not mean that household community is reducible to the same considerations that apply to social justice.
21 Consequently, Okin is misguided in thinking that the mere opposition of aims and hopes between spouses and parents makes justice, as defined in liberal theory, a crucial virtue in family life. (Okin, *Justice, Gender, and the Family*, 32.) The resolution of family conflict involves norms of a different character.
22 Walzer notes this, without however, offering any justification other than an appeal to "our" shared conventions. See Walzer, *Spheres of Justice*, 227, 229.
23 Okin, *Justice, Gender, and the Family*, 174.
24 Nozick raises this point. See Nozick, *Anarchy, State, and Utopia* (New York: Basic Books, 1974), 167.
25 Rawls, *A Theory of Justice*, 74, 105, 301.
26 Fishkin, James S., *Justice, Equal Opportunity, and the Family* (New Haven: Yale University Press, 1983).
27 Similarly, Nozick observes that families conflict with patterned distributional principles since the favoritism of family behavior incurs transfers within families that upset the favored distributional pattern. Nozick, *Anarchy, State, and Utopia*, 167.
28 Shoumatoff, Alex, *The Mountain of Names: A History of the Human Family* (New York: Simon and Schuster, 1985), 183, 188, 190.
29 For a fuller treatment of how self-determination is the locus of normativity, see Richard Dien Winfield, *Reason and Justice* (Albany: State University of New York Press, 1988), Part III.
30 For this reason, Hegel rightly claims that if one conceives ethical community as the primary topic of ethics, instead of first developing its preconditions, property and morality, one commences with the assurance of a unity grounded in nothing at all. See Hegel, *Vorlesungen Über Rechtsphilosophie*, III 480.
31 For an informative study of how contemporary Western legal systems have grappled with the issue, see Mary Ann Glendon, *The Transformation of Family Law: State, Law, and Family in the United States and Western Europe* (Chicago: University of Chicago Press, 1989), 252–290.
32 In line with the implied consolidation of private domains, Aristotle maintains that by nature complete friendship is directed at only one individual (*Nicomachean Ethics*, 1158a10–15). However, he qualifies this restriction, arguing that the limit to the number of friends consists simply in the largest number with whom one can live together (*Nicomachean Ethics*, 1171a3–7), a number that shrinks to one when an excess of friendship, passionate erotic love, is at issue (*Nicomachean Ethics*, 1171a12–14). Given how Aristotle conflates friendship with the free family, his arguments bear upon whether monogamy is the only legitimate form of marriage.
33 Aristotle, *Nicomachean Ethics*, 1155a27–28.
34 Aristotle, *Nicomachean Ethics*, 1161b12–13.
35 Aristotle, *Nicomachean Ethics*, 1160b29, 1161b18–19.

36 Aristotle, *Nicomachean Ethics*, 1160b33, 1161a2–3.
37 Aristotle, *Nicomachean Ethics*, 1161a32–35, 1161b9–10.
38 Aristotle, *Nicomachean Ethics*, 1161a3, 1161b30–1162a1, 1162a10.
39 See Plato, *Republic*, Book V, 457d ff.
40 Richard Dien Winfield, *The Just Economy* (New York: Routledge, 1988), 13–16.
41 Kant, Immanuel, *The Metaphysics of Morals*, trans. by Mary Gregor (New York: Cambridge University Press, 1991), #46, Ak. 469.
42 Kant, *Metaphysics of Morals*, Part II, Chapter 1, Section I, #25, Ak. 449–450.
43 Although palimony might seem to reduce the difference between friendship and marriage, the legitimation of palimony is open to question, especially if it applies to informal unions that are not equivalent to "common law marriages".
44 Walzer, in this connection, notes how private affairs, unlike family relations, require a "license for abandonment". See Walzer, *Spheres of Justice*, 238.
45 That the rearing of children in the family is contingent does not alter that if it does occur, it enables family membership to proceed without the mutual experience and attraction from which friendship is born.

Eleven

The Unfinished Revolution in Family Law

As the new world order makes its entry into another millennium, family values have reclaimed the spotlight that seemed ever more removed from home. At once so familiar, so unimpeachable, and yet so contested, family values have become such a fixture in public controversy that failure to question their proper boundaries and justification would be an intellectual scandal. The need for examination is all the more pressing given the prominence pervasively accorded family relations. In face of the most varied social ills and cultural depredations, the cry gets increasingly repeated that the integrity of the family is under siege and that only a reaffirmation of family values can rescue the quality of life. Politicians dare not campaign without parading their families onto the rostrum, without barraging the electorate with anecdotes of childhood care and marital commitment, without intoning how a strong family is the foundation of social and political virtue.

All this public genuflection before the gospel of family values might seem strange in light of the very developments that distinguish the modern household. For at the very moment when the family has become a central issue of public debate, more and more of humanity is experiencing a dual upheaval liberating politics from kinship relations and separating the economy from the household. With constitutional democracy and civil society approaching global hegemony, family ties maintain a diminishing hold upon political power and wealth, leaving the household a private realm distinct from the two public domains of rule and commerce. Yet, precisely by relinquishing the traditional burden of defining individuals' relation to politics and society, the family has become an independent sphere with values of its own, finally warranting determination all by themselves. Consequently, while this independence demarcates family values from social and political considerations, it sets in relief what shape the disengaged household should have, as well as how it should interact with civil society and democratic government, now that traditional institutions are fading away. These questions have a distinctly modern ring, reflecting a predicament that modern times are first making a world-wide reality. If, however, the institutions of freedom that distinguish modernity from past civilizations have normative validity, the question of the disengaged family transcends the historical moment to become a matter of determining what the family should be per se. Moreover, although not every issue of feminism and gay rights pertains to family affairs, determining the valid order of the family and how that order can be upheld without compromising the freedoms of civil society and political democracy is very much at the core of any liberation of humanity from the vestigial injustices plaguing differences in gender and sexual orientation.

The Roots of Constitutional Silence on Family Values

Since family values pertain to actual conduct and not simply to personal feelings, law cannot remain indifferent to the household so long as what the family should be has universal implications susceptible of rule-governed regulation. Further, if family values prescribe domestic relationships whose general form should never be impugned, these basic prescriptions require constitutional guarantees and not just stipulation as positive law.

Yet, as much as modern legislatures and courts face the challenge of legalizing the proper shape of the household within the framework of constitutional democracy, the most venerable modern constitution remains noticeably silent on family matters. The United States Constitution may mandate the general anatomy of a federal representative democracy, specify general areas of commercial regulation and dictate the upholding of property and contract, but otherwise our constitution holds virtually silent about how households should be ordered. As a consequence, subsequent efforts to draw the line on family relations have had to appeal to such nebulous constructions as "rights of privacy" or "equal protection" provisions, neither of which have any more applicability to family members than to single individuals.

This neglect, which parallels the Constitution's silence on economic opportunity, reflects the limitations of the liberal tradition, which has most often been relied upon to justify the new revolutionary order. If the family is to count as a sphere of right worthy of constitutional protection and legislative enforcement, these limitations must be recognized and superseded. To do so requires a certain rehearsal of well-worn moves in ethical thought.

Liberalism proceeds from the denial that reason can directly prescribe the ends of conduct, a denial itself issuing from the collapse of traditional attempts to conceive a highest good. Ethics might seem to be nothing but the search for a highest good if the worth of conduct should be measured by that of the ends it achieves. Then, the very possibility of ethics depends upon some rational ranking of ends. Since the value of ends pursued as a means for something else is relative to the worth of what they facilitate, a determination of value can only achieve finality if some end is found to have value in itself. Yet, if more than one end has a worth of its own, conduct lacks any rationale for aiming for one rather than another. Only if one end has value in itself can conduct have an unequivocal direction, and this exclusivity would require that end to figure both as an end in itself and as a master end, extending value to the pursuit of other goals only insofar as they contribute to its preeminent realization.[1] By itself, this formal requirement of a highest good threatens to license might makes right, since any prevailing rule could count as an activity sustaining itself by subordinating all other pursuits to its own will to power. Any attempt to circumvent this problem by adding further identification to the highest good, however, faces an insoluble problem of privileging some end without resting its value upon any further factor. Since introducing any other term to authorize the highest good undercuts its preeminence and since any content can be advanced immediately, reason is powerless to identify this alleged fulcrum of ethics.

Reflecting on this outcome, liberalism concludes not that ethics is condemned to futility, but that individuals are set free of obligation to any antecedently prescribed ends. Although reason's inability to privilege any goals might just as well give license to nihilism, for which freedom is no more praiseworthy than tyranny,[2] liberalism presumes that the absence of a highest good leaves liberty the only possible bearer of value, such that ends first become choice worthy by being freely chosen. Since reason cannot otherwise rank ends, individuals cannot be ascribed merit in respect to the quality of their goals. Instead, they all stand on a par as possessors of a choosing will, whose conduct has worth not because of what ends it realizes but in virtue of being autonomous. Accordingly, individuals face no other constraint than the injunction to respect the liberty of one another. This predicament allows for two basic alternatives defining the parallel routes of utilitarianism and social contract theory. If liberty be understood as a freedom of desire, with desire relegated to an empirically given personal preference, devoid of any qualitative rank, the promotion of liberty amounts to treating the desire satisfaction of each individual on a par and aiming for the maximization of the aggregate happiness of all, understood as the greatest possible satisfaction of the desires they happen to have. To the degree that these desires are merely a matter of personal preference and unsusceptible of any rational prescription, little can be said with necessity as how to best promote the aggregate happiness, let alone to make commensurable mutually exclusive desire satisfactions. Consequently, the utilitarian gambit offers no basis for any steadfast commitment to particular family norms, nor any other institutional arrangement, unless one surreptitiously introduces assumptions concerning the psychology of desire and how psychological necessity can be correlated with the results of institutional practices. The economic approach, so ascendent in jurisprudence and political science, may identify economic rationality with utilitarian principle. Yet, even markets cannot draw any unequivocal sanction from utilitarian thinking given, first, that individuals may desire matters with no price, such as political justice, the love of a beloved, artistic or scientific achievement, or philosophical wisdom, and secondly, that commodity exchange leaves consumption relative to purchasing power while determining neither the given distribution of wealth nor the unsatisfied desires with which individuals enter the market. Indeed, since representative democracy need not maximize the happiness of all, one may wonder whether anything in the Constitution finds secure grounding in utilitarian thought apart from that solitary intonation of the pursuit of happiness.

These difficulties all reflect the inability to determine universal principles for satisfying empirical desire, an inability signifying how personal preference cannot fit the form of right, which requires that whatever counts as an entitlement be something that all bearers of right can exercise without interfering with one another. To identify a liberty that can satisfy this requirement, liberalism has little alternative but to construe freedom of choice in abstraction from the content of desire. To be coherently advanced as the fundamental right of all rational agents, liberty must be construed not simply as a freedom to do as one pleases, but as a type of willing, which, no matter what end it pursues, satisfies the formal requirement of not impeding others from willing in the same manner. Since, however, the pursuit of particular ends may well conflict with the pursuit by others of their own aims, liberalism must locate a type of willing that automatically insures a mutual noninterference among those who engage in its exercise. Such a formal willing that enables agents to choose in harmony with one another has its paradigmatic embodiment in property both insofar as disposition over property involves a domain automatically demarcated from that of every other owner and insofar as respect for property abstracts from how ownership affects the advantage or disadvantage of persons. Accordingly, liberalism's embrace of formal willing treats property rights as the privileged locus of liberty, recognizing that any willing that seeks more than a freely chosen embodiment that coheres with that of others risks undermining their liberty. Social contract theory receives its baptism from this reasoning and fosters a conception of justice that essentially confines legality and the political institutions upholding law to the protection of person and property. As a consequence, any dimensions of family and social right that go beyond the disposition of property by independent individuals fall beyond the pale of justice.[3] The failure of the United States Constitution to provide any constitutional guarantees for discrete family or social rights is thus perfectly compatible with the defining limits of liberalism.

Yet can the family simply be reduced to an otherwise optional arena in which property rights apply, in the fashion of those who would remedy gender inequality by ensuring that the conditions of contractual freedom are satisfied in and outside the home? Does the family have no rights and duties of its own that transcend respect for ownership?

The rise of consensual marriage and prenuptial agreements might make it tempting to treat marriage as a contract, as if the agreement to become spouses involved disposing over external factors in which the participants have embodied their wills. Yet, how then is marriage different from any other property transaction between individuals? Should one follow Kant and distinguish marriage as a contractual exchange of the use of one another's body?[4] If this is to be more than a particular transfer of services, coexisting alongside separate private domains, the use must be exclusive and all encompassing. Then, however, the marriage agreement becomes a mutual enslavement, undermining the agencies it serves. The anomalies are no less evident when parent child relations are translated into property entitlements. Roman law may have made children the property of the father, but consistent application of that arrangement entitles parents to treat their children however they please, while leaving the passage from childhood bondage to adult independence conditional upon the parent's agreement to some sort of manumission.

The incommensurability between marriage, parenting, and property entitlements is indicative of the problem the family presents for liberalism of any stripe. Two basic features of family community stand fundamentally at odds with liberal construals of freedom and equality: first, the all-inclusive character of domestic unity, wherein independent personal property and interests are subordinated to the common good of the family, and second, the parochial partiality of family membership, by which spouses have special duties and prerogatives to one another that they do not have to outsiders, just as parents have special obligations to their own children that do not extend to other minors. On both counts, the family represents an impediment to liberal justice, constricting personal freedom by the joint sharing and caring of household union and obstructing equal opportunity and welfare in society and state by privileging the particular interest of one's home.[5] Since love, intimacy, and reproduction can all occur as much outside marriage as child rearing can be performed under public auspices, liberal considerations not just preclude constitutional guarantees for distinct family rights and duties, but encourage eliminating any juridical line between domestic ties and the impersonal, universal entitlements and obligations that apply to singles and informal cohabitants.

What gives right to the family against the incipient critique of liberalism are the same considerations that protect marriage and parenting from the Platonic charge that nepotism must subvert the common good of the state. These considerations can be encapsulated in three theses, which together provide the basis for determining valid family values and setting the agenda for the constitutional guarantee and legalization of family right:

1) self-determination has exclusive normativity;
2) the family, duly reconstructed, can comprise an institution of freedom sui generis, irreducible to the freedoms of property relations, moral accountability, civil society, or self-government;
3) the reconstruction of the family as an institution of freedom, far from conflicting with social and political justice, is a precondition for social and political freedom.

If these theses are true, conceiving the just family will amount to explicating the structure of household freedom, both in its internal constitution and in its relation to the institutions of social and political self-determination.

The Normativity of Self-Determination and the Specificity of Family Freedom

The normativity of self-determination, so central to the project of modernity and so contested by traditionalists and post-modernists alike, involves a radical break with foundations that makes the genesis of the institutions of freedom completely irrelevant to their legitimacy. Liberalism fails to carry through that break, and although the promotion of liberty first raises the banner of freedom, the freedom in question is construed as a natural capacity, whose structure is given prior to and apart from the conduct and institutions it sanctions. The faculty of choice may well be natural in that all conduct presupposes, rather than produces, the ability to choose. Self-determination, however, only occurs when individuals exercise an artificial agency, whose form is as much a product of willing as the ends it wills into being. Only then do individuals determine their own agency, achieving a self-imposed character that supervenes upon the natural identity they possess as living selves. Such self-determination cannot be achieved by a single self, for the same factor cannot independently be agent and patient, as Plato famously declared in the *Republic* to justify dividing the soul, and then the city, into different active and passive parts.[6] When, however, agents act in respect to the actions of one another, they can exercise types of agency that exist only in virtue of how their actions interrelate. This is the agency that enjoys rights, for rights are not privileges, but universal prerogatives that can be exercised only by respecting the entitlement of others to enjoy the same range of choice under the condition of honoring the correlative right of their counterparts. Self-government exhibits the self-determination realized in each structure of right perhaps most clearly since to exercise political freedom requires participating in institutions of self-rule that themselves exist in nothing but the autonomous political activity that those institutions first make possible. As participants in the conventional order of self-government, individuals engage in a form of agency that is neither given by nature nor a function of the self, but a product of the interrelated willing they therein exercise.

By attaining this self-determined character, political agency shares in the unique normativity that freedom provides. All conduct that takes its bearings from without must rely upon the validity of some external source in order to have legitimacy. But then the conduct that has normativity is different from the factor that confers normativity. This distinction between legitimating ground and legitimated entity is the stumbling block of foundationalism, which derives normativity from privileged foundations that may either be given contents or procedures of construction. The problem is that if what has legitimacy is different from what confers legitimacy, the conferring factor can never enjoy legitimacy itself. Only if what confers legitimacy issues from itself could it satisfy its own standard of normativity, but then it would cease to be the foundation of something else. Self-determination overcomes this dilemma precisely by eliminating the difference between legitimating and legitimated factors and providing conduct and institutions that owe their content to themselves. Just as what is true cannot draw its truth from something beyond truth, so what is right cannot have it measure in what lies outside right.

Accordingly, if the family can be reconstructed as an institution of right in which individuals achieve a specifically domestic form of self-determination compatible with the other modes of freedom, the household will have a normativity completely independent of any external factors. That such a household may have begun to appear within a particular epoch in a particular geographical and cultural setting cannot relativize the validity of that family order since a self-determined household has its measure in itself, rather than in any foundation.

What then are the defining features that distinguish family rights, defining what is proper to household freedom? This could appear to be a hopeless question, given how the family has traditionally been conceived and realized as an association defined by the natural necessities of procreation and nurturing to which gender differences get linked. Obviously, if the family were determined by such natural exigencies, household relations could hardly enjoy the normativity of relations of right. This does not mean that a free family cannot be an association within which sexual desire is gratified, procreation takes place, and children are raised. It simply signifies that family relations must have a character independent of the natural dimensions of these functions, a character consisting in distinct modes of self-determination.

Marriage as an Institution of Freedom

Of all the relations traditionally associated with the family, marriage appears to be the one must readily redeemable as an institution of freedom. Parent-child relations, sibling relations, and more distant kinship ties all appear rooted in biological bonds that are imposed upon individuals independently of their wills. By contrast, marriage joins individuals who hardly need have any blood relation, nor who need be wed due to any conditions extraneous to the choice of the spouses themselves. Historically, of course, marriage has been restricted by natural factors and other external constraints, such as those requiring spouses to be of different genders, banning racial, ethnic, caste, or religious intermarriage, conditioning the validity of marriage to fecundity, imposing incest taboos unrelated to upholding the autonomy of family members, or arranging marriages without the consent of the nuptial partners.

Purging marriage of all restrictions impinging upon self-determination would seem to be a simple matter: individuals would be entitled to marry whom they choose, provided the parties possess mature rational agency, enter marriage with knowledge and unconstrained resolve, and do so without having had antecedent family relations whose due integrity would be jeopardized by possibilities of subsequent marriage. Yet, if the traditional necessities of blood, rank, procreation and upbringing no longer restrict the freedom of marriage, how can marriage institute any autonomy that individuals do not already possess as single persons at liberty to enjoy sexual intimacy, child rearing, and economic and political independence apart from conjugal ties? Marriage would seem either to restrict unnecessarily the liberties of love and friendship or to represent a completely redundant formality. Upholding gay rights would accordingly require no extension of the right to marry to same sex relationships, for any such extension would add nothing to the freedoms gays already enjoy.

The dual specters of illicit restriction and redundancy disappear, however, once marriage is understood to be something beyond the pale of both contractual relations and moral accountability: namely, an ethical association, which, like civil society and representative democracy, involves community freedom where membership in the association first gives participants rights and corresponding duties whose exercise reproduces their union. Although any community imposes roles and corresponding norms upon those who belong, the normativity of freedom entails that a community is genuinely ethical when and only when its animating norms consist in nothing but rights, making membership a type of self-determined agency. What allows marriage to be an ethical community engendering rights of its own are the two characteristic features of marriage that challenge the boundaries of liberal justice, namely, the inclusivity and partiality of conjugal ties. These become the constitutive factors of a distinctly conjugal freedom when becoming spouses amounts to joining together in co-determining a common private domain into which the property and welfare of each member is merged. Such a union immediately endows its participants with rights and duties they can never enjoy simply as parties to a contract, bearers of conscience, friends or lovers, business partners, or political allies. Marriage entitles each spouse to a new freedom unattainable outside matrimony once traditional domestic hierarchies are supplanted and spouses are recognized to have the right to jointly dispose over domestic property and to have their welfare determined and cared for in common. As with any right, matrimonial freedom entails correlative duties, for members of a marriage can enjoy the entitlement to codetermine household property and have their welfare a responsibility of their partner only if all equally honor the equal right of their counterparts. Hence the joint ownership and welfare of matrimony creates its own institutional freedom only by limiting the personal pursuits and property right of each spouse. Instead of being entitled to decide *unilaterally* how property will be handled and which personally selected ends can be pursued, spouses must shape their volitions to fit the reciprocity of matrimonial right. By becoming a spouse, one's right to property is now a right to jointly manage a merged domain of ownership, just as one's welfare can now be legitimately pursued only in conjunction with promoting the ends of one's spouse. Each spouse now has a right to have his or her interest cared for by the other, just as each now has an equal claim upon the property over which the other disposes. Hence, each spouse stands in a dual relation: first, as a co-determiner of the property and welfare of the home, and secondly, as a representative of the family's particular interest in relation to outsiders. This dual empowerment may or may not increase the happiness of each spouse, but it does represent an extension of their rights to a new arena of shared privacy.

Significantly, the basic scheme of matrimonial freedom operates independently of any constitutive reference to gender, sexual orientation, or procreation and child rearing, let alone to any antecedent distinctions rooted in cultural traditions or to membership in the same social or political associations. Traditional communities that tie matrimony to social and political rank may be incompatible with matrimonial freedom, just as marriages that violate conjugal autonomy may obstruct the social and political freedom of the subordinated spouse. Nevertheless, so long as other institutions do not impose their own conflicting restraints, exercising conjugal rights requires only that prospective spouses are endowed with rational agency and already enjoy recognition as property owners and moral subjects, without which they would count as little more than slaves. Since these rights apply universally to all competent persons, irrespective of any other difference, including citizenship, they are fit for global legalization as well as constitutional guarantee. Consequently, any legal system that restricts the freedom to marry by gender, sexual orientation, nationality, religious affiliation, or any traditional hierarchies, violates right.

The need for legal reform, however, is not restricted to the right to marry; it equally applies to the freedoms that individuals first acquire within marriage. These conjugal empowerments and their corresponding duties define how the unities of property and welfare distinguishing marriage can become modes of self-determination and obtain normativity. Only by conceiving their transformation into factors of domestic freedom can the boundaries of each be rescued from the controversy that has left modern legal systems still searching for agreement regarding what rights apply to marital property and welfare.

The union of marital property has one obvious limitation: it cannot extend beyond the alienable property of spouses to their inalienable ownership of their bodies. Once the latter title is lost, one can do nothing that counts juridically as one's own. Personhood is thereby obliterated, leaving individuals in a bondage precluding participation in any further exercise of right. For this reason, spouses have no entitlement to use one another's bodies at will, as the Kantian marriage contract might suggest, and spousal battery, rape and assault remain punishable crimes.

Less obvious is whether the joint domain of marital property can coexist besides individual ownerships sheltered by prenuptial agreements, or for that matter, whether joint ownership extends only to what spouses acquire and earn during marriage, or even only to joint purchases, leaving income in separate holdings.[7]

Prenuptial agreements and other limitations upon joint marital property can serve traditional agendas at odds with domestic freedom, such as retaining property within a patrilineal or matrilineal family line rather than submitting it to the co-determination of spouses. Pre-nuptial agreements, however, more and more commonly serve to protect against divorces whose property settlements would unequally benefit the parties involved. In addition to this concern, two further non-traditional issues motivate the principal departures from communal property regimes in modern family law: the separation of assets, the community of gains, and deferred community systems. Each limits joint marital ownership at least partially to escape the unwieldiness of joint micro-management of every property transaction in marriage and partially to secure the independence of spouses by relying upon separate assets and/or income to prevent one spouse from dominating household and career decisions thanks to gender discrimination in labor markets and politics or simply due to domineering personality.

These responses address legitimate matters of marital freedom and equality. In each case, however, the restrictions upon communal ownership risk undermining the conjugal right they are intended to uphold. So long as gender discrimination persists in society and state, any separation of assets or income will only tend to increase the economic power of the privileged gender and diminish disadvantaged spouses's household co-determination and claim upon the care and resources of their mates.[8] And when gender discrimination is not a factor, any limitation upon communal ownership only insures that the more domineering spouse retains a domain completely immune from control by the more subordinate partner, as well as that inequalities in separate assets and income subvert equal marital rights within the household, especially when divorce threatens to leave spouses in very different economic straits. By contrast, when all assets and income are jointly owned, spouses cannot use differences in prior wealth and income to undermine the reciprocity of marital empowerment. Moreover, all-inclusive communal ownership need not entail impractical micro-management, since spouses can still separately manage property and careers so long as the decisions do not encroach upon the common welfare of their marriage. When such decisions do, as when transactions involve personal effects, domiciles, and major portions of family wealth or when career decisions significantly affect married life, the involvement of the other spouse becomes imperative. Of course, irreconcilable differences can always arise which threaten the unity of marriage. Nonetheless, what distinguishes the threatened unity is precisely the requirement that all property decisions that affect marriage must be subject to co-management.

If this calls into question the validity of prenuptial agreements and other restrictions upon communal property, it hardly entails divorce arrangements that violate the rights of separating spouses. Although joint ownership may leave divorcees in economic straits different from what preceded marriage, given the leveling in property that should ensue, it is debatable whether such results are pernicious, just as it questionable whether the mutual commitment of marriage can exclude risking an irrevocable shift in personal fortune.

The unification of welfare in marriage brings an added dimension to the unity of marital property. In order for the joint welfare of spouses to have validity as a type of right with its own respected freedom, the interests warranting marital care cannot be dictated by external authorities, but must either comprise conditions for marital freedom or be ends chosen by the spouses themselves satisfying the conjugal requirement of promoting their life in common as free and equal domestic partners. Although spouses alone are entitled to determine what these interests are, the chosen aims qualify as a conjugal welfare for whose attainment spouses are jointly obliged to care only if the interests neither undermine the spouses' other rights nor undermine their union as co-determining members of a joint household. Far from comprising an atomistic self-interest, subject to utilitarian imperatives, the welfare spouses are entitled to pursue and have cared for by one another must promote the reciprocally integrated interest of each.[9]

These formal parameters of the ethical welfare of marriage apply not only to the acquisition, use, and disposal of marital property, but also to all the activities that bear upon conjugal life. Admittedly, spouses' joint responsibility for maintaining one another's capabilities to exercise domestic freedom extends only to upholding the physiological and psychological health required for exercising rational agency. The latitude in the self-selection of legitimate marital interests, however, widens the scope of conjugal care to include such further factors as romance, emotional support, sexual fulfillment, and fidelity, provided spouses have agreed to make these constitutive elements in their marital welfare.

The overthrow of traditional constraints on marriage may have left individuals free to marry out of purely romantic considerations. Since, however, romantic love involves the receptivity of feeling, whose inwardness is immune from external enforcement, and ardor can accompany the most varied behavior, romance can count as a condition for entering or preserving marriage only if spouses are jointly committed to treating it as an essential part of matrimony. Similarly, with marriage liberated of any traditional requirements of procreation, spouses have no duty to engage in any type and level of sexual activity unless they have together chosen to make it an indispensable element of marital welfare. By the same token, episodes of infidelity that do not violate marital property rights need not breach duties of marital care if spouses both accept limited degrees of tenderness or faithfulness within their marriage.

This indeterminacy in the content of marital welfare extends equally to the apportionment of household chores. Although traditional gendered divisions of domestic labor violate marital duties of care by imposing tasks independently of the co-management of spouses, joint determination of marital interests empowers spouses to decide together which arrangement of household work serves their common welfare, leaving open the possibility of re-instituting traditional divisions, albeit this time predicated upon mutual consent and co-management of all household property and welfare. The same considerations govern how career decisions and willingness to rear children should be factored into duties of conjugal care. In each case, what defines the range of marital care is conformity with respect for the person of each spouse, with the other conditions of domestic autonomy, and with what spouses have otherwise co-determined to be their common interest. Of course, meeting these conditions is only possible if spouses can make their choices without being subject to constraining pressures imposed by gender discrimination in society and politics.

Since the particular content of marital welfare resists antecedent prescription, legislation must instead address violations of care and their resolution. Whereas marital law can prescribe a regime of joint ownership over all assets and income, prohibit spouses from transgressing on each other's immediate person, and require spouses to care for the health of one another, wider areas of welfare cannot be addressed without inquiring into the shared interests and expectations of spouses. For example, although a prima facie equality in household work and in social and political involvement would be desired, whether unequal apportionments violate marital welfare will depend on certifying their mutual acceptability. Similarly, lack of romance, sex, or fidelity will represent impeachments of marital welfare to the extent that they figure as factors of irreconcilable differences that shatter the unity of marriage.

The right to divorce is grounded in how fundamental disunity between spouses is tantamount to a breakdown of the ethical association that confers marital rights and duties to its members. Ethical communities of wider membership, such as civil society and the state, do not fall into desuetude when a few individuals no longer act in recognition of their ethical commitments. Because marriage extends no further than the spouses themselves, matrimony collapses as soon as they no longer treat their property and welfare as a common concern; by ceasing to perform the roles in which matrimony has its life, they render marriage an empty shell. The acknowledgment of marital breakdown does not require assignment of fault, since spouses could both bow out without malice. Nor does it require unanimity among spouses, since differences on whether to divorce can themselves reflect a chasm too deep to mend. Certifying the breach does, however, take more than testifying to a unilateral resolve, for this might comprise a transient, bridgeable difference, rather than a firm, irreconcilable opposition. Divorce law must specify the public formalities by which marital breakdown is established and withdrawal from marriage receives recognition, as well as determine how the joint property and welfare of spouses should be disassembled.

The basic principles follow simply enough from the ethical character of marriage. Other things being equal, the marital property should be divided equally since it has been jointly held. Future earnings and benefits from that future employment would seem to fall outside the reach of property settlements, since once marriage ceases, spouses no longer owe one another any special domestic care. Yet, a simple equal division of all assets at the time of divorce with no further provisions for support would itself violate the rights of marital welfare if joint decisions within marriage leave divorcing individuals with very unequal opportunities. In such cases, where, for example, a homemaker has sacrificed educational and vocational opportunities for the sake of family welfare, the resulting disadvantage should be remedied through extensions of support until an equality of opportunity is achieved.

The above sketch of how marriage can achieve normativity as an institution of freedom leaves open two questions fundamental for family ethics: 1) what if any other relations can be accommodated within family right, and 2) what modifications if any are required to insure that family freedom and social and political freedom are not in conflict with one another.

Parenting as an Institution of Freedom

Parent–child relations might seem the most resistant to incorporation in family freedom, yet it turns out that they can achieve normativity as a structure of right and that they exhaust the other possibilities for family rights supplementing marriage.

Although children, as such, stand under parental guardianship in function of their lack of maturity, this endemic subordination can figure as a relation of right provided that parental authority resides in a freely chosen commitment to bring up children to autonomy. Under these conditions, parents enjoy the right to serve as guardian of their children so long as they thereby respect their children's right to receive the care sufficient to achieve the maturity enabling them to exercise their rights as independent persons, moral subjects, prospective spouses and parents, and ultimately, as autonomous members of civil society and the state. Parents can acquire the right of guardian through procreation and if they are responsible for the birth of a child, they are prima facie obliged to provide for its upbringing to autonomy. Since, however, legitimate parenting consists in an ethical relationship, it cannot depend upon blood ties. Just as parents forfeit custody of their offspring if they violate their parental duties, so qualified individuals can elect to adopt children, irrespective of gender, sexual orientation, ethnicity, religion, citizenship, or any other factor that does not impede providing for the development to autonomy of the children. Whether parents and children be biological kin or not, the rights and duties remain the same: parents must furnish children with whatever care is needed for physical and psychological maturation, together with the education enabling them to exercise all their rights. Children, for their part, are subject to due parental discipline, but are not obliged to perform any domestic services that go beyond what contributes to their own autonomous development. When children reach maturity, the juridical parent-child relation eliminates itself insofar as children achieve recognition as independent persons, liberated from the guardianship of their erstwhile parents, who themselves become freed of the responsibility of parental care. Although parents and grown-up children may retain special emotional ties, these cease to impose any specific rights and obligations beyond those that apply to moral subjects in general. Accordingly, adult children have no domestic claims to the property and support of their parents any more than parents have domestic rights to be cared for by their children. Indeed, since the regime of communal property obviates inheritance between spouses, given how marital property remains owned by the survivor, family relations entail no rights of inheritance beyond the retention of family wealth sufficient to finance the upbringing of surviving minor children.

Beyond marriage and divorce, family right extends no further than these restricted boundaries that parent-child relations assume once they become configured as structures of freedom. Although siblings may share parental tutelage as minors, their very immaturity leaves them without any domestic rights and duties to one another during their upbringing. Upon maturity, their sibling relation has no juridical claim upon any subsequent family memberships they assume as spouse and parent, except for incest exclusions that could be justified by the need to prevent the veil of household privacy from subjecting children to sexual abuse.

Relatives at any further remove, such as grandparents, aunts and uncles, cousins, or nephews and nieces, not to mention step-relatives of every variety, are no less bereft of any family claims to the property or care of related spouses, parents or children. Since all these extended relations do not belong to the shared private domain of spouses or of parents and children, they cannot enjoy the specifically domestic rights and duties that membership in household community confers. Consequently, although the nuclear family should be liberated of gender discrimination and heterosexual limitation, family right cannot accommodate the impositions of traditional extended households, where spouses, parents and children find themselves subordinate to other relatives.

Another "New Birth of Freedom"

The outline of family right sets an agenda for constitutional supplement and positive legislation, without which the freedoms of self-government remain incomplete. American democracy needed to abolish slavery, eliminate property qualifications, extend the vote to women, and tear down Jim Crow to begin to realize the "new birth of freedom" that Lincoln welcomed. Yet all these steps remain half-measures so long as the rights of social and family freedom are not extended to all. The New Deal and The Great Society attempted to address some of the obstacles that obstruct equal social opportunity and enable social disadvantage to translate into political inequality. Yet just as social rights remain without constitutional guarantee, so family rights still remain largely homeless in the American Republic. Inequality and oppression in the home cannot help but obstruct the universal exercise of social and political freedom, since whomever is lorded over at home is thereby hindered from entering society and politics on an equal footing. Consequently, the precarious plight of family right leaves government of, by, and for the people a promise awaiting future fulfillment. Just as economic relations must be modified to enable individuals to exercise equal political freedom, so society and politics must undergo their own modifications to insure that individuals are not socially and politically handicapped by their household obligations as spouses and parents. Addressing this challenge depends, of course, upon first establishing the proper boundaries of family values. Determining family right, however, only sets the stage for considering how social and political self-determination can be made compatible with the freedoms of domestic community. Altering the schedule of work, commerce, schools, courts, and political activity to fit the rhythms of domestic duties, guaranteeing affordable child and spousal care, and preventing career advancement from being jeopardized by domestic responsibilities are all too familiar challenges awaiting solution. Only when individuals are guaranteed the meeting of such challenges will they be free at home, at work, and in the political arena, putting the lie to Plato's worry that what is good for the family cannot be good for the state.

*****

Although family freedom comprises an enabling condition for equal economic opportunity, economic autonomy has a character of its own, distinguishing it from both household and political self-determination. The call for economic democracy both challenges and sets in relief the differentiation of these different spheres of right. The following chapter accordingly investigates whether applying democratic decision-making to the economy promotes either social or political freedom.

Notes

1 Aristotle gives this argument its classic statement in Chapter 1, Book I of the *Nicomachean Ethics*.

2 Michael J. Sandel notes this lacuna in liberal thought in *Democracy's Discontent: America in Search of a Public Philosophy* (Cambridge, MA: Harvard University Press, 1996), 8.

3 Thinkers, such as Fichte and Rawls, who have employed social contract arguments to ground more inclusive welfare considerations fail to recognize how the normative imperative of providing individuals with the conditions enabling them to achieve universalizable ends calls for no more than protecting person and property. For a detailed discussion of this failure, see Richard Dien Winfield, *The Just Economy* (New York: Routledge, 1988), 34–54.

4 See Kant, Immanuel, *The Metaphysics of Morals*, trans. By Mary Gregor (New York: Cambridge University Press, 1991), sections 22–23, AK 276–277.

5 Accordingly, such leading contemporary liberal theorists as John Rawls and James Fishkin see the autonomy of the family implacably impeding the realization of distributive justice. See John Rawls, *A Theory of Justice* (Cambridge, MA: Harvard University Press, 1971), pp. 74, 105 & 301, and James S. Fishkin, *Justice, Equal Opportunity, and the Family* (New Haven: Yale University Press, 1983).

6 See Plato, *The Republic*, Book IV, 430e–431a.

7 For an informative survey of how modern legal systems have differed in drawing the boundaries of matrimonial property, see Mary Ann Glendon, *The Transformation of Family Law: State, Law, and Family in the United States and Western Europe* (Chicago: University of Chicago Press, 1989), 116–140.

8 See Glendon, *The Transformation of Family Law*, 134–5, for a discussion of this problem.

9 For this reason, the attempt to determine family values by applying the utilitarian, allegedly "economic", calculus of atomistic self-interest, such as undertaken by Gary Stanley Becker in his *A Treatise on the Family* (Cambridge, MA: Harvard University Press, 1991) and by Richard A. Posner in his *Sex and Reason* (Cambridge, MA: Harvard University Press, 1992), is thoroughly misguided.

Twelve

Should the Economy be Democratized?

The Call for Economic Democracy

The collapse of communism may have cast a pall upon the attractions of central planning and state ownership of production, but it has not silenced the call for economic democracy. With the transition from Marxist–Leninist rule to political democracy and market economy raising anew the problem of relating civil society and the state, might the extension of democracy to the economic arena be the specter haunting the new world order? Yet why democratize the economy? Political justice may require democratic government, but does either political freedom or social justice call for applying democratic decision-making to economic affairs? Economic democracy has been advanced on both accounts with various rationales.

On the one hand, advocates have argued that political justice cannot be fully achieved without democratizing the economy. Some, such as Norberto Bobbio, have maintained that the very future of democracy lies in extending democratic rule beyond the political domain to overcome the limits upon popular participation endemic to modern representative government.[1] Others, such as Charles Lindblom,[2] have proposed economic democracy as necessary to remedy the subordination of politics to privileged groups of civil society. Finally, some, such as Carole Pateman,[3] have advanced economic democracy as a needed training ground for political involvement, providing a mediation between the particular interests of civil society and the universal aims of politics. On the other hand, a democratic economy has been proposed as an imperative of social justice. For certain advocates, such as Michael Walzer[4] and council communists, economic democracy is a necessary ingredient in economic freedom, without which economic oppression prevails. Moreover, for them and others, democratizing the economy is just as necessary to curtail economic inequality.

Complicating the debate over these imperatives are the varied forms that economic democracy can take. Worker self-management comprises one basic option, yet it alone can assume very different guises according to the scope of its jurisdiction, the way it integrates management and ownership, and how it is mandated. The scope of worker self-management can be limited to an individual work place or extend to an individual corporation or to a whole industry or cartel. Further, no matter how extensive be the enterprise it controls, worker self-management can further vary according to how much a prerogative it enjoys over the needs, objects of satisfaction and work assignments of its members, and to whether or not its jurisdiction extends to the municipal concerns of a self-governing commune. Moreover, worker self-management may combine control with exclusive ownership or manage an enterprise partly or wholly owned by other private interests or the state. In addition, all of these possible arrangements can be established through the voluntary agreement of those involved or mandated by law as the exclusive form of economic administration. Alternately, economic democracy can dispense with worker self-management entirely and instead directly administer the economy by the decisions of representative government, be it under a centralized national regime or under a federal system.

Needless to say, any question of whether economic democracy should be mandatory or optional only pertains to economic democracy as a form of enterprise management. When economic democracy is exercised as a direct administration of the whole economy by parliamentary institutions it is just as mandatory as any government office. However, if the democratization of the economy is conceived as worker self-management, it will have to be legally mandated if it is required by any imperatives of social or political justice. Leaving its institution voluntary will not suffice, for worker self-management can hardly be guaranteed to arise *laissez-faire* as the exclusive form of enterprise organization. This is especially true if self-management is combined with ownership. Then the enterprise will be left at a competitive disadvantage by its inability to raise additional capital by such market mechanisms as the issuing of bonds and shares.

Given all these possibilities and their implications, can any form of economic democracy claim to be an imperative of social or political justice?

Can Economic Democracy Remedy Social Disadvantage?

To begin with, can economic democracy remedy social disadvantage, let alone be a necessary instrument for doing so? A major stumbling block to this vocation lies in the arbitrariness of democratic decision-making.[5] At the level of individual enterprises, the mere institution of worker self-management hardly ensures that the economic opportunity of all its own members will be equally promoted. The majority may exploit the minority or impose any variety of arrangements no more equitable than those of firms managed differently. Of course, this license could be held in restraint by legal limits upon what can be democratically decided concerning working conditions, pay, benefits, promotion and the like. However, what then guarantees equal economic opportunity is not worker self-management but the external regulation imposed either at the political level through the enforcement of legislation or at the civil level by public welfare institutions.

A similar predicament prevails if the economy is directly administered by representative institutions. Here again, unless the arbitrariness of majority decisions in parliament is curtailed by constitutional limitations, the enforcement of equal economic opportunity cannot be guaranteed. If constitutional statutes do succeed in that regard, it is their mandate and not the democratic form of the collective management of the economy that precludes economic disadvantage.

On the other hand, even if the members of a self-managed enterprise dealt each other an even hand, their democratic decisions within their enterprise need not promote the economic welfare of those outside its bounds. Not only may a worker self-managed enterprise benefit its members at the expense of its direct competitors, but it may fail to provide affordable commodities meeting the needs of other consumers. Moreover, no firm's internal decisions can unilaterally guarantee that its members will not end up disadvantaged compared to individuals in other enterprises in the same or other branches of industry. Since economic democracy has meaning on the enterprise level only if it leaves each enterprise an autonomous market agent, having to accommodate other independent commodity owners in order to satisfy its own needs, none can guarantee positive ramifications of its decision-making for its own members or any one else. Each is just as much captive as collaborator to the unforeseen play of market forces as any other kind of enterprise.

Similar limitations afflict the ability of worker self-management to provide for other matters of economic right, such as public infrastructure facilitating the general exercise of economic freedom or protection of the public health and environment. Nor can economic democracy be counted on to prevent social privilege from translating itself into political privilege by enabling particular social groups to amass a preponderant share of the resources for political action.

Is Economic Democracy a Necessary Ingredient in Economic Autonomy?

Yet, if economic democracy cannot remedy economic disadvantage, can it still be a necessary ingredient in economic autonomy?

Two types of oppression have been ascribed to market relations in the absence of economic democracy. The first, bewailed by Marx and every other objector to market interdependence, is the subordination of all market participants to an economic necessity of which none is the author. Such subordination to an economic order whose law is consciously willed by no one is endemic to market activity and, indeed, to all action in civil society. Because individuals, as participants in the market, pursue particular ends and never will the universal order to which they belong, their economic activity can never exhibit the reflexivity characteristic of democratic political action, whereby citizens not only will the universal order of the body politic, but do so by acting upon their own activity, imposing law upon the community to which they belong. It is precisely this unwilled character of the economic order that makes possible an objective economic science, since otherwise, the law of the economy would be a matter of convention, knowable only by a historical cognition that studies the history of enacted economic arrangements.

In addition to this tyranny of an anonymous economic necessity, which is the common curse of all members of the market, a second, more selective oppression is duly lamented. This consists in the subordination of some individuals to the will of certain other market agents on whom they are economically dependent. Indeed, such submission is also endemic to market activity, with the qualification that *all* participants must accommodate themselves to the needs of others in order to satisfy their own. But is this subordination a restriction of freedom or rather a constitutive element in the exercise of economic autonomy, involving the same mutual limitation that enters into any exercise of right, so far as enjoying a right entails observing the duty of respecting the same right of others? Or does it reduce equal economic opportunity to a purely formal reciprocity, where the options of the economically dependent are illusions masking the one-sided bondage of wage slaves? In any event, can economic democracy liberate individuals from either alleged oppression?

To begin with, so long as markets are retained, it does not matter whether individual enterprises, whole corporations, or entire branches of industry are worker self-managed. However the internal organization and ownership of firms be arranged, they and their members will automatically be subject to an economic order that they never will as the object of their economic activity. That submission to the invisible hand of market lawfulness can only be overcome if economic democracy is exercised at a national level through a total central planning that replaces markets with a unilateral public allocation system. In that case, the "economic" order gets reduced to the stipulation of a technical agency, acting upon the economy as an object to be manipulated. However, by replacing the market interaction of self-determining agents by a technical administration of things, such economic democracy ends up canceling any specifically economic autonomy. For although individuals may have an equal voice in the collective determination of central planning, they all lose the right to choose independently what they will need and what activity they will pursue to satisfy their needs in reciprocity with others. Moreover, in losing this right, citizens gain no enhancement of their political autonomy, since they can exercise the same rule over the economic domain without sacrificing their economic freedom by democratically determining how markets should be publicly regulated. Hence, the escape from the social necessity of markets turns out to eliminate the very autonomy it seeks to promote.

Economic democracy fares no better as a remedy to the subordination of individuals to other particular economic agents. If the targeted oppression consists in a one-sided economic dependency that renders formal the equality of commodity exchange, no democratization of enterprises can eliminate the problem. Whether the dependency be rooted in disparities of wealth or in the monopolization of certain economic factors needed by others, worker self-management cannot overcome it simply because the democratization of an enterprise still leaves it dependent upon other market agents with whom it must trade to survive. Hence, a democratized enterprise can just as well lord over other financially weaker firms and individuals as be lorded over by its stronger counterparts. Admittedly, inequalities in wealth and monopoly power can be restricted by market regulation, thereby extending equal economic opportunity to self-managed enterprises and their members. However, then it is not enterprise democratization that overcomes economic dependency, but a market regulation that can address the underlying problem without needing to impose self-management.

What, however, about the unilateral prerogatives exercised by the owners of enterprises over their employees, be these owners private or public? Do these not comprise an oppression that enterprise democratization can alone overcome? If, as we have seen, economic inequality and monopoly power can be curtailed by market regulation so that the relation between employer and employee need not be a purely formal agreement, masking one-sided domination, what oppression is it that worker self-management addresses? When Michael Walzer condemns private ownership of capital for entailing the tyranny exercised by Pullman over his company town, Walzer fails to recognize that the conflation of municipal authority with private ownership once characterizing Pullman, Pennsylvania, has nothing to do with the strictly economic prerogative that ownership of an enterprise confers.[6] The example of Pullman rather represents a private enterprise that has extended its control beyond the work place to municipal governance, transgressing the divide between civil society and the state.

Worker self-management certainly can operate in conformity with market and political freedom and thereby enjoy legitimacy as one option among others in which market agents have the right to join. Nevertheless, even when enterprise democracy gives employees control over their work place, that control will effect the economic options of others in the market who do not participate in this management. Once again, public regulation can counteract the impact of these ramifications of market interdependence, but this signifies that public regulation can do the job apart from enterprise democracy. Moreover, if self-managing workers have exclusive ownership of their enterprise, this limits their ability to raise capital, introducing a further impediment to enhancing the economic opportunities of those within and without the individual enterprise.

Alternately, if the rule of enterprise owners over their employees is counteracted by supplanting the independence of enterprises with a democratically controlled central planning, the independent economic autonomy of each citizen is equally eliminated, without adding to their opportunity to participate in politics.

By contrast, maintaining the market with employment separated from ownership and control of enterprises does not automatically undermine either economic or political autonomy. Although employees will then submit to the work place rule of their employer through voluntary contract, by itself this submission does not violate their autonomy as an economic agent any more than it undercuts their freedom as a citizen to collectively regulate the entire economy. So long as public regulation upholds equal economic opportunity, the absence of mandatory worker self-management does not interfere with exercising the economic right to pursue personal needs through activity of one's choosing in reciprocity with others. By the same token, so long as public regulation prevents economic factors from subverting equal political opportunity, citizens can still have full access to the freedoms of self-government, no matter what their relation to the ownership and management of their enterprise. In this way, a class society need not foster class rule, where representative government becomes "bourgeois democracy", indirectly securing for enterprise ownership the political domination that Pullman's municipal company rule directly expressed.

Is Economic Democracy a Remedy to Civil Society's Subordination of Politics to Particular Privileged Interests?

Since economic democracy by itself does not eliminate economic disadvantage, it can hardly prevent the subordination of politics to privileged groups that rests upon economic inequality. However, in the context of civil society and the state, economic disadvantage is not identical to political disadvantage.

On the one hand, persistent economic inequality may still be prevented from infecting the political domain if the resources for political action are strictly regulated so as to prevent economic privilege from affecting the political options of individuals. What makes the difference is not economic democracy, but such strictly political measures as public funding of political parties and campaigns, public access to the media, and restrictions upon private financing of political activity.

On the other hand, the regulation of economic inequality to achieve equal economic opportunity may still allow for political disadvantage. Although inequalities of wealth may be limited so as to enable all individuals to exercise their right to earn a conventional standard of living through activity of their own choosing, differences in resources may persist that give some privileged access to political action. This can occur simply in virtue of certain individuals' economic role. For example, owners and employees of communications and transportation industries can use their economic position to influence the conduct of political campaigns to a degree that others cannot match. In such cases, economic democracy is of no avail, since it may well be the workers of a self-managed enterprise who use their economic role to gain disproportionate political influence. Only a strict public regulation can counteract these possibilities.

Is Economic Democracy a Training Ground for Political Participation?

Nevertheless, economic democracy might still have an abiding value as a key training ground for preparing citizens for political involvement. This view, advanced by Carole Pateman,[7] has its forerunner in Hegel's conception of the corporations of civil society.[8] By offering individuals a chance to unite in conscious promotion of a common aim, these social interest groups were supposed to provide a bridge between the personal pursuit of particular ends emblematic of civil society and the conscious pursuit of the universal order of the body politic characterizing political activity.

Hegel's suggestion is fundamentally flawed, however, because the interest consciously advanced by a social interest group may be common to its members, but still remains a particular interest lacking the universal character of political ends. Consequently, "corporate" advocacy lacks the two cardinal features of political action: it is not universal in aim, nor is it fully self-reflexive in the way in which political agency imposes upon itself the order of the community in which it exercises self-rule. Hegel obscured these limitations of corporate activity by illicitly importing into civil society a universal class of civil servants whose task consists in implementing the universal aims of government.[9] This move, of course, heralds Marx' identification of the proletariat as revolutionary subject in virtue of being a class whose interest is universal.[10] However, the proletariat can have a universal interest only if it be considered completely outside civil society's pursuit of particular ends, enjoying the same externality shared by Hegel's universal estate of civil servants, whose role is completely extraneous to the economic functions by which classes are defined. That, however, contradicts its very own character as a class of civil society.

Consequently, economic self-management cannot serve as a training ground for political participation because of the fundamental distinction between action in civil society and political activity. Economic self-management could only serve that tutelary purpose if politics were itself reduced to a national housekeeping, losing its own status as an association for the sake of its own activity of self-rule.

That reduction could only be accomplished by collapsing the distinction between state and civil society, placing the economy under an erstwhile "political" central administration. Since this would eliminate civil society itself, the end result would be an undermining of the two pillars of public freedom.

Contrary to the admonitions of Pateman et al, political freedom presupposes a liberation from the confines of social interest. This liberation is not to be found through the self-assertion of any common particular interests but rather through the activity of politics itself, whereby it not only upholds pre-political freedoms as preconditions of political autonomy but insures that their exercise does not compromise self-government. To be sovereign over civil society and genuinely self-governing, the state must mediate its own relation with civil society and the family, rather than allowing itself to be based upon some social factor.[11]

Can Economic Democracy Remedy the Limits of Representative Government?

Since economic democracy concerns the pursuit of particular ends generic to civil society, it does not involve properly political action. Therefore, economic democracy cannot serve to mediate between the individual political will of the citizen and the general will of the state. Although worker self-management may afford individuals a say in their work place regarding their own economic activity, it in no way extends their participation in political self-determination. For that, citizens must engage in specifically political institutions, such as political parties organized not around work place, industry, class or any other strictly social unities, but around different political programs.

Nor, for that matter, does economic democracy add in any way to citizens' political control over the economy. That prerogative, which involves the dual imperative of regulating the market to insure equal economic opportunity and to insure equal political opportunity, is exercised by participating in the legislative, authorizing and executive functions of government that presides over economic activity. Participation in these political functions operates irrespectively of how enterprises are owned and managed.

The only way in which "economic democracy" might be thought to add to political freedom is if the prerogatives of economic self-management were extended to political affairs. This could occur either by self-managing enterprises transforming themselves into self-ruled communes or by democratically controlled central planning being exercised in conjunction with collective political rule. In either case, however, economic freedom is eviscerated without any gain in political autonomy.

Indeed, if parliamentary government is transformed into a council system of self-managed enterprises two problems are introduced. One is the subordination of politics to economic management, a reduction consciously sought in the reveries of instrumental reason guiding council communism. At one blow, that subsumption of society and state to a collective "administration of things"[12] sacrifices social and political autonomy at the altar of a technical mastery of nature.

On the other hand, if this limitation is removed by defining a council system apart from work place and economic management, constitutionality is undermined.

Hannah Arendt proposes a pyramid of councils of this sort, whose members are self-selected at the local level and then elect members to the next ascending council whose members do the same for that above them, and so on up to the highest pinnacle of authority.[13] Although this frees the council system from the taint of class interest and a reduction to a national housekeeping, it prevents any upholding of the division of powers on which constitutionality rests. The different levels of governance can be distinguished according to their degree of local jurisdiction, but at each level, the same group of council members exercise all functions of government. Lacking is any independent authority to hold action in check within each level or to adjudicate between jurisdictional disputes between levels.

To uphold constitutionality and insure a commitment to the different modes of freedom, two conditions must be met that elude the political extensions of economic democracy. To begin with, politics must function as something for its own sake. If instead the state is legitimate by realizing pre-political aims, then it operates as a means to an end. Consequently, the mere existence of the state cannot guarantee its fulfillment of its legitimating purpose, which is defined apart from the actuality of the state. Therefore a problem automatically arises as to how to compel political institutions to realize their non-political ends. No political institution can possibly accomplish that mission, since each such institution will remain a means to an end given apart from its own activity.

The quandaries of liberal political theory exhibit this dilemma. Because liberal theorists erect their civil government as a means to enforce a liberty defined prior to and apart from political institutions, they are left with two options, each of which is self-defeating. On the one hand, they must recognize that the sovereign can do no wrong since its activity cannot be limited by any independently effective law or institutional counterweight. Thus Hobbes will argue that the sovereign cannot be bound by law, since law is nothing but the command of the sovereign, who can always rescind his previous edicts,[14] just as Rousseau will argue that the general will can never do wrong since it issues from and applies to all citizens equally,[15] echoing Hobbes' own claim that the Sovereign cannot commit injustice since its subjects are the author of its will.[16] Yet, by thereby admitting the absence of any institutional resource for controlling government, they supply the rationale for a Locke to allow for a right to rebellion based on the personal judgment of individuals,[17] sapping civil government of any abiding recognition and reintroducing a state of nature.

By contrast, if the state is for its own sake, and not an instrument for realizing a pre-political liberty, then so long as it exists, it achieves its legitimating aim, enjoying a self-sufficient, self-grounded authority. This, however, requires a separation of a legislative branch that formulates law, an authorizing power that makes these formulations valid, and an executive branch that enforces the law through government administration and judicial proceedings. When each of these branches is distinguished, yet determined by some form of participation of all citizens, then a constitutional order can be maintained where no individual or particular group can substitute for the self-rule of citizens.

Any council system that reduces government to a means to economic management and/or unifies all powers at each level cannot provide these prerequisites for constitutionality and political self-determination. On this score as well, economic democracy is no remedy to the perceived formality of political autonomy in representative government.

The Common Mistake of Socialism and Economic Democracy

What then is to be done with the call for economic democracy? Like the call for socialism, which it sometimes accompanies, it arises from a fundamental confusion. Paralleling socialism's rationale for nationalizing commerce, economic democracy mistakenly roots economic disadvantage in the form of ownership and control of enterprises and in the corresponding class divisions. Yet as the history of the various socialist experiments continues to confirm, the problems of equal economic opportunity are not to be solved by any alteration in the internal form of ownership and management of enterprises, nor do class divisions automatically bring with them any specific advantage and disadvantage once the full spectrum of commodity relations are taken into account.[18] Neither the nationalization of the economy nor a universal proletarization guarantees any particular degree of equal economic opportunity, nor any particular realization of any other mode of freedom. All that can be said for sure is that state ownership establishes the most powerful monopoly power of all, with all its attendant dangers for social and political freedom, just as a universal proletarization can be counted on to limit the economic options from which individuals are free to choose. Instead, the focus of attention must shift to the public regulation of market institutions, a regulation which must always be called upon to make good what is perennially left unresolved by alterations in forms of ownership and management.

Just as socialism seeks a remedy to economic disadvantage in alterations of ownership and class that hinder economic freedom without achieving equal economic opportunity, so economic democracy turns to a form of management that is equally unable to realize social and political justice. Both panaceas divert attention from the genuine problems afflicting civil society and the state.[19] Whether they are acknowledged or not, these problems remain on the top of the agenda of the new world order, posing the dual challenge of regulating markets so as to insure equal economic opportunity and to prevent economic factors from undermining the autonomy of politics and the equal political opportunity of citizens.[20]

*****

If the shortcomings of economic democracy draw attention to the relation between civil society and the state, they leave unimpeached the promise of political democracy. Yet if political democracy is to be upheld, the legitimacy of self-government must be emancipated from the relativism that has increasingly been associated with it. The next chapter calls into question that linkage, so as to secure the justification of democratic government.

Notes

1 Norberto Bobbio, *The Future of Democracy*, trans. by Roger Griffin (Minneapolis: University of Minnesota Press, 1987), p. 62.
2 Charles E. Lindblom, *Politics and Markets* (New York: Basic Books, 1977), Chapter 24, p. 330–343.
3 Carole Pateman, *Participation and Democratic Theory* (New York: Cambridge University Press, 1970), p. 110.
4 Michael Walzer, *Spheres of Justice* (New York: Basic Books, 1983), pp. 292–303.
5 This arbitrariness equally applies to democratic decision-making at the political level of government. Hence, abstractly speaking, political democracy does not of itself guarantee the upholding of social justice, whereby the enforcement of equal economic opportunity enables members of society to enjoy their economic freedoms. However, social inequality can impede citizens from exercising equal political freedom, both because political action requires economic resources of its own and because relations of economic dependency can prevent individuals from taking advantage of their political rights. Consequently, the realization of political freedom is not independent of the realization of a just society in which individuals escape bondage in their economic, or, for that matter, household relations. Indeed, it is just because the upholding of family and social freedom is a precondition for political self-determination that political freedom is not undercut by constitutional requirements compelling government to respect the household and social rights of its citizens.
6 Walzer, *op. cit.*, pp. 297–303.
7 Carole Pateman, *op. cit.*, p. 110.
8 G. W. F. Hegel, *Philosophy of Right*, trans. by T.M. Knox (New York: Oxford University Press, 1967), paragraphs 202–205, 250–256; pp. 131–132, 152–155.
9 For an extended critique of Hegel's concept of classes, see Richard Dien Winfield, *The Just Economy* (New York: Routledge, 1988), pp. 149–156.
10 See Karl Marx, "Contribution to the Critique of Hegel's Philosophy of Right – Introduction", in Karl Marx, *Early Writings*, translated and edited by T.B. Bottomore (New York: McGraw Hill, 1964), pp. 55–59.
11 For a further discussion of this political mediation, see Richard Dien Winfield, *Reason and Justice* (Albany: State University of New York Press, 1988), pp. 251–258.
12 Arendt notes how Engels puts the matter in this telling fashion. See Hannah Arendt, *On Revolution* (New York: Viking Press, 1976), p. 276.
13 See Arendt, *op. cit.*, pp. 282–284.
14 Thomas Hobbes, *Leviathan*, edited by C.B. Macpherson (Harmondsworth, UK: Penguin Books, 1968) pp. 234, 313, 367.
15 Jean-Jacques Rousseau, *The Social Contract and Discourse on the Origin of Inequality*, edited by L.G. Crocker (New York: Pocket Books, 1967), Book II Chapter 3 of *The Social Contract*, pp. 30–32.
16 Hobbes, *op. cit.*, p. 232.
17 John Locke, *The Second Treatise of Government*, edited by Thomas P. Peardon (Indianapolis: Bobbs-Merrill, 1952), Chapter XIX, p. 138 ff.
18 For a detailed discussion of how this is so, see Winfield, *The Just Economy*, Chapters 6–8.
19 For a further discussion of these abiding problems as they afflict contemporary regimes, see Richard Dien Winfield, *Overcoming Foundations: Studies in Systematic Philosophy* (New York: Columbia University Press, 1989), Chapter 9, pp. 171–216; and Richard Dien Winfield, *Freedom and Modernity* (Albany: State University of New York Press, 1991), Chapter 15, pp. 261–282.

20 For a discussion of the principles of such public regulation, see Richard Dien Winfield, *The Just Economy*, Chapters 10 and 11, pp. 183–232. It should be noted that whereas public regulation of markets is needed to uphold family, economic and political freedom, entailing the specific equalities of opportunity that apply within these three institutional spheres, these imperatives do not involve any mandate to foster "economic democracy". As the preceding arguments have shown, "economic democracy" is at best an optional arrangement providing no necessary contribution to the constitutive exercise of household, social or political self-determination.

Thirteen

Relativism and Democracy

Relativism and democracy have long been regarded as corollary theory and practice. The alleged connection has not always favored majority rule, but increasingly, the tie has been advanced as the key for justifying democracy. The saving relativism gets variously identified – as a liberal neutrality of ends, an indifferent bazaar of ethical communities, or a pragmatic accommodation to contingent privileged vocabularies. However defined, the privileged relativity is held to empower majority rule. Although the new world order might seem to have placed democracy beyond contention, the resurgence of ethnic nationalisms and traditional fundamentalisms serves notice that the legitimation of democratic government cannot be taken for granted. Consequently, the turn to relativism as a justification for democracy must be called into question.

The Classical Rejection of Democracy as Relativism

The very idea of legitimating democracy by appealing to relativism seems hopelessly absurd in light of the arguments that led Plato and Aristotle to dismiss majority rule. Indeed, Plato famously pointed out that what deprives democracy of validity is precisely the relativism that democratic rule entails. By treating all citizens as political equals, whose opinions are to be given the same weight in determining the norms ruling the community, democracy not only devalues all differences in knowledge and character, but leaves it a matter of indifference what ends are made the common good. Since all that counts is whatever pleases the majority, democracy abandons reason for appetite, removing politics from ethics and eliminating any possibility of a rational justification of government.[1]

So long as reason mandates any ends apart from majority rule and its preconditions, the Platonic argument would appear to be irrefutable: by privileging nothing but its own choice procedure, democracy relativizes the ends to which rational government should be firmly committed.

Yet by condemning democracy for relativism, the Platonic refutation presents a negative requirement for the legitimation of democracy: majority rule may be justifiable only if reason cannot prescribe any ends of government apart from self-rule and its prerequisites. This might seem to suggest that the justification of democracy does depend upon relativism, that only if reason is unable to prescribe a highest good, can majority rule command legitimacy.

Aristotle, of course, had argued that unless some end could have an absolute, rather than a relative value, all conduct would be in vain, since any rationale for action would fall into an infinite regress in which no goal could be found worth achieving in its own right.[2] Could such a predicament possibly give license to democracy?

Democracy and the Relativism of Ends

Since Hobbes, liberal thinkers have assumed that the inability of justifying a highest good left liberty the sole factor of conduct possessing legitimacy. Admittedly, the absence of a highest good may well free agents of any obligation to achieve ends prescribed antecedently to their choice. Since, however, the lack of prescribed ends equally leaves the door open to a nihilism for which liberty is no better than oppression, it hardly follows that conduct has validity only insofar as it is freely chosen.

The latter assumption has been employed by some, but not all, liberals to legitimate democracy as a corollary of liberty and of the relativism of ends thought to make freedom of choice the exclusive bearer of normativity in conduct. Rousseau, Kant and, more recently, Rawls and Habermas have all argued that majority rule follows from the "priority of right over the good", i.e. the premise that no candidate for highest good has an exclusive legitimacy and that liberty should be advanced as the privileged determiner of valid conduct and institutions.

Even if we ignore the challenge of nihilism and grant the exclusive legitimacy of liberty, it is far from self-evident that liberalism can provide an adequate argument to privilege democracy over other forms of government. Although liberalism has historically been most frequently appealed to for ideological support by democratic movements, thinkers from Hobbes and Locke to Isaiah Berlin[3] and Hannah Arendt[4] have recognized that upholding liberty with the consent of the governed is not identical to institutionalizing self-government. Hobbes[5] and Locke[6] both realized that a civil government protecting the person and property of individuals could have the consent of the governed, as given through social contract, and still allow for different forms of sovereignty where one, the few or the many actually rule. Because liberty simply involves freedom of choice, defined in function of an agent not yet engaged in any legitimate institutions, it need not involve any participation in collective decision-making.

In order to be a right, however, liberty must be a form of choice that can be exercised without conflicting with the analogous choosing of others. For this reason, liberty cannot be coherently interpreted in utilitarian fashion as a freedom of desire, of doing as one pleases. Not only is the satisfaction of one desire potentially at odds with the satisfaction of another desire of the same person, but nothing guarantees any harmony between the desires of individuals, let alone, any objective way of establishing commensurability between desire satisfaction that might make a utilitarian calculus possible. For these reasons, those advocates of liberty who recognize the incoherence of a utilitarian interpretation of rights turn to property ownership as a paradigmatic realization of liberty. Unlike democratic participation, property can realize liberty as a right that individuals can possess prior to any other normative institutions. Political freedom, like social, family and moral autonomy presupposes respect for property rights insofar as individuals must be recognized as at least owners of their own bodies if they are to escape enslavement and be able to exercise any further rights at all. Although moral initiative and legal institutions may be needed to ensure that property rights are universally upheld, individuals can determine themselves as owners simply by giving their wills recognizable, mutually exclusive embodiments and by respecting one another's complementary domains. For this recognition process to proceed, the participants need not belong to any further associations, such as families, civil society, or the state. Since ownership involves an exercise of choice that automatically coexists with the disposal over property of every other owner, *without* presupposing membership in any community, it provides a construal of liberty that can serve as a legitimating principle for public institutions without falling into a vicious circularity. Fatal circularity would result if liberty were instead defined as an institutional freedom, for then the principle by which institutions are determined to be just would presuppose the legitimacy of a particular institutional arrangement. The advocates of liberty can escape this problem, as well as the dilemmas of utilitarianism, by determining justice as both an outcome of an exercise of ownership, i.e. contract, and as a realization of property rights under public guarantee. This governing recipe of social contract thought, however, leaves democracy at best a tolerable option among the forms of civil government. Although democracy might serve liberal ends, democracy need be no more committed to upholding the property rights of all individuals than need be monarchy or oligarchy. The many, after all, may desire the property of the few.

Both Rousseau and Kant nevertheless privileged democracy on the grounds that liberty could not be realized by government unless the will of the state, rooted in the legislative power, was co-determined by all citizens. Only then could individuals be subject to a law they have imposed upon themselves, instead of being ruled by some particular interest. In other words, only democracy in legislation would allow government to realize freedom and transfigure whatever ends it pursued into universal aims. Rousseau himself raised doubts as to whether collective decision-making could be counted upon to realize liberty or to impose ends of sufficient universality: unless all citizens chose the same end, the majority would only represent a particular interest, and even if all made law unanimously it might have a content violating right. Hence, not only must all political factions and social roots of particular interests be repressed, but a civil religion and a legislator must be imported to tame and train the general will.[7]

Rawls and Habermas have attempted to redeem the justification of democracy from Rousseau's equivocations, reformulating the relativism of ends in terms of cultural pluralism and a relativity of reasonable notions of the good. The basic argument retains a familiar structure. Given the assumption that what is to be decided in political process cannot be unequivocally mandated by reason alone, respect for liberty translates into a deliberative democracy, where each individual's view should have equal sway. What counts, politically speaking, is whatever emanates from the deliberations of collective determination.

If, following Habermas, one identifies the framework of deliberative democracy with the conditions of rational discourse, rationality gets caught in a peculiar bind. With rational justification procedurally defined, what is rational becomes inscrutable until democratic deliberation has run its course. But in that case, what can democratic *deliberation* involve? The situation is comparable to the predicament that legal realism imposes upon judicial decision-making. According to this view, law has no objective significance apart from what judges decide. Hence, as Oliver Wendel Holmes famously declares, law is really but a prediction of judicial behavior. Yet, if that is so, what are judges supposed to be doing in arriving at their judgments? If the law has no prior objective determination, what argument can they make to defend whatever decision they hand down? And how can they predict their own decisions if no antecedent rational principles can serve as grounds? By the same token, what can a Habermas offer as a content for democratic deliberation to render it something other than a contest for supremacy of certain irrational interests over other equally irrational interests? It would seem that procedural democracy ends up ratifying Carl Schmitt's claim that conditions of mass democracy and party discipline rob parliament of any reality as a deliberative body.[8]

If deliberative reason is so formal that none of its results can be known prior to the polling of all reasoners under conditions of unconstrained discourse, the very theory of deliberative democracy must be suspect. Both the "argument" determining the proper form of democratic deliberation and the "argument" privileging that form as the deciding procedure for determining valid norms can hardly be the outcome of the deliberative process. Yet if the theory *of* deliberative democracy is not determined by collective choice, it makes appeal to a reasoning that *can* operate independently of the alleged conditions of rational discourse, subverting its own claims.

The Rawlsian idiom of the original position suffers from the same problem, whether in its first statement or in the guise of political liberalism. The alleged fact of reasonable, yet unresolvable controversy over ethical value is supposed to leave individuals with an equal right to determine the principles of justice and political institutions. Yet what is "reasonable" must already be limited by an acceptance of equal rights, since a relativism of the good hardly translates into a commitment to numerical equality. Then, however, the specification of the original position and of the choice procedure it privileges rests either upon arbitrary assumptions (including an acceptance of the alleged "moral intuitions" of our modern community) or upon a reasoning that can determine norms independently of procedural determinations.

Moreover, the Rawlsian argument for democracy is at pains to legitimate the conclusion that collective decision-making will best achieve the realization of each individual's self-selected ends. Regulated markets, for example, provide an institution that enables individuals to satisfy self-selected needs in reciprocity with others, whereas a collective administration of the economy would cancel that form of autonomy. Political participation need not be one of the aims of personal choice, nor need majority rule secure more self-selected goals of citizens than a benevolent despot. Because liberalism proceeds from a conception of choice that is a function of a disengaged agent, rather than an institutional freedom embedded within a specific form of community, the relation between liberty and institutions remains fundamentally indeterminate.

Democracy and the Relativism of Ethical Communities

The communitarian turn seeks to overcome this indeterminacy by recognizing how the form of a community determines the type of agency, ends and governing norms of its members, providing a common framework within which values can have an intersubjective, albeit historically relative, validity. Because communitarians conceive community as a formal idea that does not specify the particular agency, ends and norms it incorporates, these are regarded as historical givens, that can at best be heuristically clarified rather than impartially evaluated. The formal idea of community can therefore hardly provide a justification of democracy as a universally valid form of government. Instead, the relativism of ethical communities suggests that democracy can be "legitimated" from within the contingent horizon of whatever particular culture may already associate its members in terms of democratic values. The justification of democracy is thereby set free of the relativism of ends; now democracy will have legitimacy in virtue of sustaining a form of community that will count as an end in itself for its members.

This eliminates one of the fundamental stumbling blocks to any justification of democracy: the difficulty of legitimating democracy as a means to an independently given end. So long as democracy is thought of as justifiable only for instrumental reasons, the constitutive caprice of majority decision will pose a daunting problem, for which no restraining measures can provide a coherent solution. If the will of the majority is accorded sovereignty, there is no telling what ends it will adopt, whereas if majority decision is constrained to ensure the fulfillment of certain privileged goals, democratic freedom is undercut by competing concerns.

Yet can communitarian relativism provide a genuine rational for treating democracy as an end in itself? The "civic virtue" of democratic community may be embedded in the roles that membership involves, but can such virtue have genuine authority for its citizens if the legitimacy of democracy rests upon the contingent existence of the community? The communitarian is aware that liberty does not command exclusive legitimacy simply because a highest good cannot be justified by reason; the specter of nihilism must not be forgotten. Yet, is nihilism really overcome by accepting the communitarian conclusion that, in face of the inability of reason to mandate ultimate ends and the inability of liberty to have justification by default, the only alternative to *subjective* values are the *intersubjectively* recognized norms internal to historically given communities? Such communities may impose obligations upon their members, but if these obligations have no reach beyond the contingent existence of the community, how can they claim authority against those who would overthrow the community to establish a different one in its place? As soon as the existence of the community is called into question, which, given its contingency, is always possible, its internal norms loose their hold upon individuals, who remain capable of adopting a new form of association. Although reason may offer no impartial grounds for preferring one community for another, the grounds for retaining allegiance to the one to which one belongs are themselves arbitrary. On this basis, membership in a democratic community cannot give democracy more than a qualified status as an end in itself: it has non-instrumental value only *on the condition* that one abide by its own internal norms. Communitarians, such as Sandel, may well be able to interpret the contemporary American scene as a democratic community whose civic virtues need strengthening in order to comprise a consistent whole.[9] Similarly, Partha Chatterjee may turn to elements of Indian tradition to identify a non-Western, yet modern post-colonial democracy in which collective self-rule has internal validity.[10] Yet so long as such efforts retain adherence to communitarian relativism, where legitimacy is conditioned by historical conventions, any normativity they claim for their respective democracies remains merely hypothetical. Such hypothetical normativity will not only leave enemies of democracy unswayed, but it will burden self-critical democratic citizens with the awareness that their most fundamental allegiance is irrational.

Democracy and the Relativism of Contemporary Pragmatism

Contemporary pragmatists, led by Richard Rorty, would like to believe that the self-critical awareness of the contingency of community, and indeed of all justification (e.g. "privileged vocabularies"), comprises an irony for which democracy is the most suitable expression. Unlike other political formations, democracy allegedly embodies a neutrality to doctrines of human nature, as well as to truth in general, so long as truth is understood to involve more than general acceptance by a contingently given community. This neutrality is the counterpart of pragmatism's ironic stance to philosophical, that is, absolute claims of knowledge and value. Instead of demanding some foundation in nature or reason on which to legitimate democracy, the new pragmatism claims that democracy has no foundations and needs no philosophical justification precisely by comprising a political framework that restricts itself to weighing opinion, leaving foundational issues to private reflection. The new pragmatists tells us that because we can never engage in justification without appealing to a privileged vocabulary embedded in some contingent practice, we should forego the demand for ultimate justification and restrict ourselves to making our practices more coherent. But can we be coherent in restricting ourselves to such an enterprise if our practices themselves raise issues of justice, not as hypothetical, pragmatic issues, but as categorical principles of right? Does a democratic society alone allow for being coherent in this manner, insofar as democracy limits its deliberations to matters of opinion? If the new pragmatism is to accord any special warrant to democracy, it would seem to rest on an affirmative answer.[11]

Yet even these reflections cast a cloud over any special legitimacy for democracy in general or for the mere viability of democracy as an option for our modern community. The new pragmatism may continually deny that it is pursuing a narrative about theory and practice with its own privileged vocabulary, but if it is not engaging in a meta-discourse, how can it disqualify the possibility of other alternatives and be confident of its own irony? Moreover, what can forestall irony from embracing a will to power, which opposes democracy with a new aristocracy, brazenly affirming the arbitrary particularity of its aims as the only strategy consistently recognizing the contingency of value? Rorty assures us that the Nietzschean option will strike the democratic ironist as a madness, afflicting individuals who may retain the capability for communicating with us, but are incapable of coherently belonging to our democratic society, whose internal consistency is the only measure we have.[12] Yet if we recognize the authority of democracy with due irony, as dependent upon the historical accident of our membership within a democratic society, nothing prevents us from reflecting on the predicament this implies for truth and value and opting for the fascism that a consistent will to power fosters.

Democracy without Relativism

Despite these problems, the new pragmatism turns out to be right on one crucial point: that democracy needs no foundations, and indeed cannot have any. What the new pragmatism fatally misunderstands is that freedom from foundations is not equivalent to the contingency of norms, from which irony issues.

To the degree that democracy is a structure of political freedom, it exhibits the liberation from external determination that distinguishes all forms of *self*-determination. A democratic polity does not owe its character to any independently given nature or normative standard: imposing its own order upon itself, democracy is its own ground. In this respect, democracy is an end in itself, for if it were instrumental, it would not set its own ends, but be predetermined by a given function.[13]

As a consequence, democracy can escape the pitfalls of all institutions that are justified by appeal to instrumental considerations, or more generally, by appeal to foundations. Foundational justification falls prey to the very dilemma that freedom overcomes: the distinction between what confers validity and what possesses validity. A foundation that legitimates what is founded can never satisfy its own standard of normativity, which ascribes legitimacy only to what derives from the privileged foundation. For the foundation to meet this requirement, it would have to be its own foundation and thereby cease to be the ground of something other than itself. In other words, the foundation would have to be self-determined to escape incoherence. In effect, self-determination resolves the dilemmas of foundational justification. By eliminating the difference between legitimating standard and legitimated factor, freedom commands a normativity that anti-foundationalism never even begins to consider.

Although the liberation of democracy from foundations might seem to absolutize majority rule to the detriment of any other normative commitment, it invites neither utter arbitrariness nor an elimination of non-political rights. Self-determination is not devoid of structure. Far from being indeterminate, democratic freedom consists in the workings of very specific political institutions, involving elected representation, division of powers and constitutionality.[14] Further, although the *existence* of freedom may well depend upon certain natural, psychological and historical conditions, its structure, qua self-determined, is not defined by any given particular conditions. Freedom therefore has a universality allowing for its comprehension by philosophical conception. Moreover, insofar as political freedom requires that pre-political institutions, such as property relations, the family, and civil society, not impose restrictions that interfere with the equal political participation of citizens, democracy must uphold the pre-political rights of its citizens in order to sustain their political rights. If it can be shown that the preconditions of political freedom consist in according citizens their rights as property owners and members of a free family and civil society, then democracy will have legitimacy both as an end in itself and in upholding the other freedoms of citizens. The "proceduralism" of majority rule is therefore not formal, for its own operation will depend upon respecting and protecting self-determination in the other spheres of right. Although these imperatives are internal to democratic freedom, they are not relative to the historical existence of a particular community. Instead, they are dictated by the universal structure of self-determination. Although democracy does free citizens from bondage to any ends other than those sustaining self-government and its preconditions, it does not usher in the hegemony of opinion. Once democracy is duly recognized as an institution of freedom, and the special normativity of self-determination is comprehended, the connection between relativism and democracy, assumed by its traditional detractors and contemporary advocates alike, falls by the wayside.

The implications for modernity should be evident. To the degree that modernity is distinguished from other forms of civilization by realizing the institutions of freedom, modernity will have no more foundations than democracy. The inability of the Enlightenment Project to find foundations for the modern age is therefore anything but a strike against modernity.[15] So long as modernity is characterized by the universal structure of self-determination, it, like democracy, will overcome relativism and command a normativity that neither foundationalists nor anti-foundationalists can successfully undermine.

*****

The legitimacy of democracy and of the other institutions of freedom endow the project of modernity with a normative significance decisive for comprehending the tensions unfolding between "developed" and "underdeveloped" nations and within "developing" nations themselves. The following chapter explores these abiding oppositions by investigating the legitimation problems of the different paths to modernization that define the post-colonial predicament.

Notes

1 Plato, *Republic*, Book VIII, 557 ff.
2 Aristotle, *Nicomachean Ethics*, Book I, 1094a20–21.
3 Isaiah Berlin, *Four Essays on Liberty* (Oxford: Oxford University Press, 1969), 129–130.
4 Hannah Arendt, *On Revolution* (New York: Viking Press, 1976), 168–170.
5 Thomas Hobbes, *Leviathan* (Cambridge: Cambridge University Press, 1991), Chapter XIX, 129–138.
6 John Locke, *The Second Treatise of Government* (Indianapolis: Bobbs-Merrill, 1952), Chapter X, 73–4.
7 Jean-Jacques Rousseau, *On the Social Contract* (Indianapolis: Hackett, 1987), Book II, Chapter III, 31–32, Chapter VII, 38–41; Book IV, Chapter VIII, 96–103.
8 Carl Schmitt, *The Crisis of Parliamentary Democracy*, trans. by Ellen Kennedy (Cambridge: MIT Press, 1988).
9 Michael J. Sandel, *Democracy's Discontent: America in Search of a Public Philosophy* (Cambridge: Harvard University Press, 1996).
10 Partha Chatterjee, *The Nation and Its Fragments: Colonial and Postcolonial Histories* (Princeton: Princeton University Press, 1993), Chapter 11, 220–239.
11 Richard Rorty, *Contingency, Irony and Solidarity* (New York: Cambridge University Press, 1989), 68–9.

12 Richard Rorty, *Objectivity, Relativism, and Truth: Philosophical Papers Volume 1* (New York: Cambridge University Press, 1991), 187–8, 190.

13 Although classical political philosophy duly conceived the just state as an end in itself, by still construing political justice in the functional categories of embodied form, classical philosophers like Plato and Aristotle failed to recognize how only what is self-determined can be an end in itself.

14 For arguments supporting these points, see Richard Dien Winfield, *Reason and Justice* (Albany, NY: State University of New York Press, 1988), 270–282.

15 For a probing development of this point, see William Maker, *Philosophy without Foundations: Rethinking Hegel* (Albany: State University of New York Press, 1994), 21–46.

Fourteen

Post-Colonialism and Right

The Problem of Post-Colonialism

Amidst the euphoria greeting the end of the Cold War, the temptation has been great to welcome a new world order heralding the global triumph of democracy and free enterprise with a human face. Following fascism's defeat a half century before, the sudden collapse of super power rivalry may relieve much of the totalitarian challenge to civil society and constitutional self-government. Yet celebrating the completed march of reason in history remains premature. Not only do daunting problems still plague the transition from communism to capitalism as well as capitalist democracy itself, but another unresolved opposition demands ever more attention. While the world stage has been commandeered by a century of inter-imperialist, fascist/anti-fascist, and communist/capitalist slaughter, the great majority of humanity has undergone a different ordeal. Traced by overlapping rubrics of colonialism, Westernization, modernization, and development, and culminating in the political liberations creating the post-colonial condition, this parallel history can no longer lie outside the limelight.

Both as an external opposition between North and South, the developed and the underdeveloped, or the West and its other, and as an internal opposition within post-colonial society, a chapter of history is unfolding, distinctly different from the birth of the modern that left and right Hegelians once viewed as the final combat. How is this development to be understood? Can the categories with which the master thinkers of modernity comprehended our age grasp the post-colonial predicament?

The problem at issue has key descriptive elements, most important of which concerns the contingent or non-contingent character of the contrast between how Western nations modernized themselves and how non-Western nations were generally subjected to a "Westernization" imposed through the yoke of colonial and imperial domination. Above all, however, what lies at stake is prescriptive inquiry addressing first the normativity of modernity and of the Western developments initially giving rise to modern institutions, but then extending to probe the separability of modernization and Westernization, the legitimacy of imposing modern institutions from without, and finally the normative tensions within post-colonial societies and between developed and underdeveloped nations.

All of these normative questions are decided as much by how modernity is understood, as by how normativity is itself conceived. Defenders and detractors alike generally agree that modernity distinguishes itself from prior forms of civilization by calling into question given tradition and demanding that practices and institutions command legitimacy only to the degree that they are justified by reason. Hence, the institutions that modernity erects putatively in accord with reason lay claim to a universality reflecting their independence of the contingent particulars of given authority. Not surprisingly, the distinctively modern institutions that are understood to be universally valid are institutions of freedom, determined not by their conformity to particular cultural tradition, but by their realization of self-determination. The institutions giving reality to self-determination can qualify as the real that is rational precisely because freedom exhibits the independence from foundations that reason must exhibit if it is to provide any justification that does not rely upon dogmatically accepted criteria, rules, procedures of construction, or any other privileged vocabularies. Precisely because free institutions do not draw their legitimacy from any conformity to given tradition, they can only be justified by a reasoning that rejects the traditional view that justification amounts to derivation from some foundation. Thus, when MacIntyre decries the failure of the Enlightenment to establish foundations for modern autonomy,[1] he is not exposing a crack in the facade of modernity, nor unmasking limits in reason; rather, as William Maker has so powerfully argued,[2] the very failure to locate foundations for modernity is indicative of how the universality of modernity's institutions of freedom involves their independence from all prior grounds.

It is the universality of modernity, enshrined in institutions of freedom, that allows the modern to represent not just a particular moment in history, inevitably overtaken by the post-modern of a later date, but a uniquely valid form of civilization, valid in the special sense of not falling prey to the problems of legitimacy that afflict any practices that claim authority on the basis of privileged foundations.

Traditionalist opponents of modernity can certainly cling to their fundamentalisms, especially when self-professed modern regimes fail to realize freedom in all its proprietary, moral, household, social and political dimensions. So long, however, as the traditionalists' opposition to modernity rests upon an appeal to given sacred or profane authority, they cannot overcome the arbitrariness of their own norms, norms that define pre-modernity to the degree that modern institutions emerge through the overthrow of foundational tradition.

Post-modernists, on the other hand, recognize the futility of absolutizing any particular foundations, but they assume that no values and no knowledge claims can be advanced without depending upon some privileged vocabulary or other. On this assumption, modernity's claim to enjoy universal validity and freedom from the given is bogus, concealing some ground reflecting a particular culture or group whose advance of universal norms can only be a play for power, imposing its own values upon all. This post-modern challenge to the legitimacy of modernity has one basic difficulty: if all normative claims are foundational, leaving reason and conduct incapable of self-determination, that is, free of determination by prior grounds, the truth of this very situation can never be authoritatively established. Post-modernists must either know with an autonomy they deny or admit that their claims are as ideological as any others. Hence, as much as post-modernism may give ideological support to the fascist challenge to modern institutions of freedom, it cannot refute the project of modernity.

Although modernity may enjoy special normativity, the traditional, modern, and post-modern alternatives represent the abiding, unmixed options that a civilization can take with regard to foundations: i.e. it can remain traditional, privileging a particular set of given practices, it can be modern, embracing institutions of freedom, or it can be post-modern, advancing a particular form of life with brazen recognition of its own perspectival, arbitrary character. Accordingly, once all three forms of civilization have made their appearance, it is possible to speak of an "end" of history in the restricted sense that history may offer repetitions and hybrid combinations of these options, but no fundamentally novel forms of community.

Where post-colonial society fits in such a scheme results in large part from two consequences following from the universality of modernity.[3]

The Dual Character of Emergent Modernity

First of all, because modernity's institutions of freedom do not depend upon any particular culture for their legitimacy, they are inherently capable of global, not to mention, inter-galactic, realization. By contrast, because pre-modern civilizations are distinguished by appealing to some particular heritage as their defining basis, they cannot consistently break with their parochial roots and achieve cosmopolitan hegemony.

Secondly, although the universality of modernity renders it capable of globalization, nothing in the structure of the institutions of freedom requires, guarantees or even makes possible that they come into being universally at once. Rather, the supplanting of traditional forms of life by the different spheres of modern community is bound to occur originally in some particular locality marked by some particular heritage. Moreover, wherever modernity originates, its distinct freedoms can only emerge in a series of developments fostering some of the institutions of freedom before other types come into being. The impossibility of simultaneous birth is due to the structural relation of the different modes of freedom. Self-government may preside over all other forms of freedom, upholding and regulating their integration within the body politic. Yet no political act can freely create property relations, moral autonomy, freedom in the family, or a civil society. This is precluded because citizens cannot engage in self-government unless they already count as property owners, moral agents, and equal partners in marriage, while enjoying equal economic opportunity and legal rights.[4] A deficit in any of these autonomies undercuts the equal political opportunity of citizens. This is why genuine political democracy need not be externally constrained to safeguard the pre-political rights of citizens. On the contrary, self-government cannot actualize itself unless citizens already enjoy their pre-political autonomy in family and society, whereas democracy cannot sustain itself unless the state continues to uphold these pre-political spheres of freedom. Accordingly, the genesis of political emancipation must be preceded by all the domestic, social, and cultural developments that make possible household and social emancipation, developments whose own time and geographical extension are prey to all the vagaries of historical accident that preclude either an instantaneous or uniform formation. Hence, even if the final release from feudalism may at one blow disengage kinship from commerce and rule, while separating social position from political power, the revolution that baptizes a free family, a civil society, and a republic will still present the universal rights of modernity in a regional debut.

Taken together, the universality of modernity and the regionality of the birth of the modern render the globalization of modernity something more complicated than merely a transformation of the pre-modern into the modern.

To begin with, the advent of modern institutions in a particular region creates a divide between two forms of pre-modernity. One pre-modernity stands distinguished as the womb of original revolution – that is, as that pre-modern civilization that transforms itself into the modern. The metamorphosis of this pre-modernity has a pathway defined by what must occur for the institutions of freedom to arise. The requirements of this original genesis of the modern is of course dictated by the structures of freedom and the reformations of culture that make it compatible with the exercise of rights. Accordingly, the conception of this genesis is posterior to the conception of the institutions of right, which is why Hegel in the *Philosophy of Right* is correct to analyze the history of the emergence of freedom[5] only after having conceived property right, morality, family, civil society, state, and international relations. Nonetheless, even if the genesis of the modern must include specific transformations, this requirement does not dictate that the historical emergence of modernity is itself necessary.

Hegel, of course, has often been alleged to treat the history of freedom, that is, the genesis of modernity, as a necessary process. Significantly, however, his lectures on the philosophy of history begin with the empirical *hypothesis* that institutions of freedom have in large part arisen in our day, a hypothesis which then enables us to interpret history up to the present as a history of the emergence of freedom.[6] Moreover, in analyzing that history, the different way stations do not comprise successive transformations of the same community. Rather, Hegel gives us more of a transmigration of soul, where the development of freedom leaps from one nation to another, albeit within the general orbit of the West, before a modernity arises that can spread its wings beyond a particular people and make itself a global civilization. If each successive form necessarily generated the next stage, one would expect a metamorphosis of one and the same people. That this is not the case suggests that although the original genesis of modernity may have a form dictated by its conclusion, the arrival at that destination rests upon much that is contingent.

Even if this is so, once modern institutions first arise within a particular region, all other communities stand in a new situation. Some may manage, at least temporarily, to close themselves off from contact with the first modern society and then autonomously metamorphose into a modern society in their own right (as Japan may have done), or alternately linger in their own isolated pre-modernity. Wherever such seclusion is overcome, however, pre-modern societies become subject to a modernization from without. Since the original modern society has a regional identity, the modernization it may foist on other pre-modern societies will take on the appearance of a regional assimilation, even if it involves the establishment of universally valid institutions. In terrestrial terms, insofar as the original modern society has arisen in the West, the external modernization will appear in some respect as a process of Westernization.

Whether "Westernization" and "modernization" completely converge depends upon several factors. If the transformations in question simply entail establishing purely universal structures of freedom, "Westernization" will be identical with modernization from without. If, alternately, the transformations involve imposing practices specific to Western culture in its contingent particularity, "Westernization" becomes a particular form of modernization that may well diverge from the modernization that an isolated non-Western community may autonomously achieve. The latter option, assumes, of course, that what distinguishes Western civilization are not cultural features that uniquely enable a community to modernize itself autonomously. If non-Western culture did lack such features, as Hegel, Marx and Weber all suggest, then the only way non-Western civilization could be modernized would be under the aegis of an external Westernization. In that case, a rigid divide would separate the pre-modernity that transformed itself into Western modernity and the pre-modernity of non-Western civilization.

The Globalizing Tendencies of Emergent Modernity

Whether or not a pre-modern community is able to modernize itself, the whole question of an external modernization remains irrepressible due to the globalizing tendencies of any emergent modernity.[7] These globalizing tendencies have two parallel dimensions, one rooted in the structure of civil society, the other following from the exclusive normativity of self-determination.

Although civil society comprises an ethical community whose members interact in function of the pursuit of particular self-selected ends, their association is universal in scope, reaching as far as market interdependence, legal recognition, and economic welfare considerations extend. Political community may erect frontiers separating different states, but civil relations transcend all such borders. This is why economic, legal and other civil rights warrant international guarantee under the rubric of "human rights", applying equally to non-citizens and citizens, in contrast to national rights to political participation.[8] It is also why states must take special measures to prevent transnational social developments, such as the growth of multinational enterprise and multinational financial and regulatory integration (e.g. the EEC), from undermining national sovereignty. Moreover, as Hegel early observed,[9] and Marx and Rosa Luxemburg[10] would later stress, the commodity relations of civil society drive it beyond itself to penetrate pre-modern societies in search of an expanding market furnishing new consumers, new sources of raw materials and manufacture, and new pools of labor. With civil society endowing individuals with the freedom to determine their needs and occupation under the social condition of enabling others to do the same, the resulting market demand engenders a multiplication and refinement of needs and production, liberating the accumulation of wealth of any natural limits.[11] The resulting pressures of competition make an enterprise's survival depend upon increasing sales and profits and the investment these make possible. To the degree that competitive success cannot always be achieved by increasing market share at the expense of other firms or by maintaining market share in a growing internal market, the expansion of civil society's market into other regions becomes an imperative of its own economic dynamic.

How this break through occurs is colored by modernity's general normative tendencies to globalization. On its own self-understanding, the modern community must regard all pre-modern formations as oppressive regimes violating the rights that only institutions of freedom can actualize. Even though traditional communities must command sufficient consent to maintain their own authority and may well be enthusiastically supported by the overwhelming majority of their members, voluntary conformity to tradition is not equivalent to self-determination. This discrepancy between preference and freedom is most obvious in the case of popular dictators: they may have the consent of the majority, but they cannot provide the participation in self-government that political self-determination requires. The same deficit applies to the voluntary acceptance of family arrangements ruled by traditional hierarchies or of social orders organized by hereditary rank: without the institutions of free households and civil society, individuals cannot enjoy their family and social rights even if they experience domestic and social happiness. To the degree that modernity judges itself by the foundation-free standard of freedom, it must regard right as what alone has universal validity and not as a parochial heritage, in the manner of Rawlsian advocates of "reflective equilibrium" and communitarian particularists. But then modernity cannot consistently celebrate the diversity of pre-modern civilization; modernity must instead regard global modernization as a normative imperative.

This imperative would apply to pre-modern civilizations whether or not they have the capability for an internal, autonomous modernization. If a contemporaneous traditional community lacked the inner resources to revolutionize itself, then the modern community would face the burden of engineering that transformation from without. If, alternately, a traditional community had the ability to modernize itself, the co-existing modern community would still face the challenge of facilitating that metamorphosis with whatever external aid is possible.

These imperatives of global modernization assume that self-determination can be externally imposed, or, more specifically, that modernization can be forced upon pre-modern civilization without corrupting the desired establishment of right. Yet can communities be compelled to be free? It is tempting to answer this question by drawing an analogy between the problem of making another individual autonomous and making another community self-determined. Kant formulates the general problem in his *Doctrine of Virtue* by arguing that individuals cannot make others autonomous, e.g. moral, but can only promote their happiness by helping secure the satisfaction of their desires.[12] Kant's argument might appear to apply uniquely to moral autonomy since moral self-determination revolves around choosing to act with the right purposes and intentions, an internal matter that external intervention can hardly control. By contrast, the other freedoms exercised in property, family, economic, legal and political relations have an external dimension that could be affected from without. After all, if determining oneself as an owner, a spouse and parent, a market agent, a legal subject, and a citizen depends upon recognition by others and common exercise of the freedoms comprising the different institutions of right, self-determination will depend upon the contributions of other agents. Nevertheless, Kant argues that colonial or imperial domination of another people is unjust no matter what norms the latter obey and no matter what the liberating intentions of their new masters.[13] Since Kant models right after the self-legislation of moral autonomy, any external imposition of institutions will automatically lack that self-legislating character, even if the new order establishes the putative structures of freedom.

Yet might the absence of self-legislation in the establishment of the institutions of freedom be endemic to the process of modernization, whether that process be internal or external in origin? Self-legislation as an institutional engagement is specific to democratic self-government. Spouses and parents may codetermine household management, but the scope of the family is a particular right and welfare to which legislation cannot be restricted.[14] Although market activity may conform to economic law, economic agents do not consciously enact the rules of competition. Only self-government involves a freedom with the universal reach requiring rule by law. Nevertheless, as a structure of ethical community, political democracy can itself only operate on the basis of an existing constitution, whose own establishment is not an act of self-government, but rather the result of a founding process involving a plurality of antecedent developments. As Hegel has emphasized,[15] the emergence of self-government depends upon the formation of pre-political institutions, including the universal recognition of property rights (and the abolition of slavery), the acceptance of moral autonomy (and the religious reformation it requires), the rise of civil society's market economy (and the overthrow of feudal bondage), the institution of civil courts and due process, and the formation of public welfare agencies to guarantee equal economic opportunity. Only with these pre-political transformations, can any founding of a democratic constitution signify more than the charade that "people's democracies" once represented. Since these enabling pre-political developments are no more pre-ordained than is political revolution, actual constitution making is, as Hegel points out, a contingent affair of history, fundamentally different from self-legislation.[16] This is true whether the constitution making be an internal affair or a product of external modernization.

Either way, the interconnection of pre-political and political freedom indicates that modernization requires a preponderant engagement of individuals in each of the practices in which right has its exercise. Moreover, because these practices are modes of self-determination, modern institutions cannot be imposed from without unless the recipients are willing to break with their interrupted tradition and interact in terms of their received rights. Although such willingness cannot be reduced to self-legislation, it has its own internal dimension.

Accordingly, it makes little sense to follow the Kantian/Rousseauian gambit of judging internal and external modernizations by whether they result from self-legislation. A more plausible strategy consists in first conceiving how the structure of the different institutions of freedom determines the possible interactions between an original modernity and the pre-modern communities it confronts. Once these possibilities are identified, a better judgment can be made of the ethical problems of external modernization and of the colonial and post-colonial predicaments it engenders.

Modernity and the Logic of Colonialism

To understand colonialism and its post-colonial aftermath in relation to modernization, one must think through how the two globalizing tendencies of modernity work themselves out once a modern society has arisen in a particular locality. Since the modern norms that call into question pre-modern communities equally preside over how the emergent modern society interacts with its pre-modern counterparts, it makes sense to examine first how the globalizing dynamic of civil society impacts upon pre-modernity and then to investigate how the principles of modern freedom apply to the resulting situation.

Although Hegel's analysis of the market is notoriously sketchy, he does identify the salient features that push civil society beyond itself, setting the stage for colonial and imperialist expansion. Enabling individuals to relate in terms of the pursuit of self-selected needs for commodities, the market not only multiplies and endlessly differentiates needs, means of satisfaction, and the forms of earning, creating demand for commodities that lie beyond its own borders, but equally engenders a predicament where demand may fall short of supply, leaving civil society too poor to sustain independently its ever expanding pursuit of wealth.[17]

This confronts civil society with complementary alternatives to realize the economic freedom of its members: On the one hand, if civil society cannot adequately reduce unemployment through internal public interventions such as non-redundant public works or subsidized incentives for employers, it must export its own unemployable labor. On the other hand, if sufficient internal market growth cannot be sustained by credit expansion and deficit spending, civil society must draw upon new supply and demand for which the only remaining source is the abiding pre-modern civilizations that have still escaped modernization.[18]

These parallel injunctions might seem achievable by means of the basic modality of civil association: commodity exchange. Although pre-modern society as such limits civil freedom by traditional divisions that organize household, economic, and legal relations by factors given independently of the self-determination of individuals, commodity exchange can still conceivably play a circumscribed role within the interstices of traditional community. Trade between pre-modern and modern society is therefore possible, as the history of early European penetration of the non-Western world can testify. Unemployed settlers can find earning opportunities in the markets of pre-modern society, just as modern enterprises can obtain labor power, raw materials, and finished goods in exchange for whatever commodities (including money) traditional communities are willing to obtain in return. Yet can such possibilities satisfy the economic dynamic of civil society, or, for that matter, enable a traditional community to modernize through the purely economic expansion of market relations within its borders?

What frustrates either of these prospects is the fundamental disparity between the limits pre-modern tradition places upon economic freedom and the unlimited pursuit of wealth that an unshackled market generates. Merchants and workshops in pre-modern society may earn through exchange with a fully developed civil society. Nevertheless, their own level of trade and production is inevitably constrained by the traditional arrangements that impose fixed limits upon the needs and occupations of themselves and their compatriots, subordinating their own commerce within a wider scheme of social reproduction from which commodity relations are excluded. Because of such traditional limits upon what can be freely bought and sold, market activity is not only constrained in ways that threaten the competitiveness of indigenous industry, but incapable of autonomously assimilating the economic relations that are regulated by non-market pre-modern relations.

The consequences of this limitation are prefigured in the genesis of the original civil society. Marx documents the key factor in his analysis of the "primitive accumulation of capital",[19] expressly contradicting the base-superstructure economic determinism that would enfeeble Marxism. Namely, the only way that market relations can expand their reach in pre-modern society is through a political intervention that forcibly uproots future wage laborers from the conditions of their traditional sustenance and compels them to seek their living in the market. Otherwise whatever wealth has been already accumulated through trade and usury will be unable to find continually expanding opportunities for investment and fail to precipitate a transformation of the economy into a system producing goods for exchange and the realization of profit.

What applies to the genesis of the first civil society extends with a special twist to external modernization. Whereas primitive accumulation occurs through annihilation of the traditional livelihoods of people by political actions of their own rulers, lifting the barriers to market penetration in other pre-modern societies from without can only occur through an imperial domination where the government of an emergent modern society uses its state power to force a pre-modern community to "open" itself to the globalization of the market. This "opening" cannot simply consist in allowing foreign goods to be imported for purchase; any such trade concessions still leave intact the pre-modern social formations that exclude commodity relations. Instead, those traditional arrangements must be uprooted by political means. The globalization of civil society through external modernization will thus require an external political intervention, enlisting intrigue and military force to overcome indigenous resistance. As the example of China indicates from the Opium Wars through the adventures of the Comintern, such intervention can retain local rulers but force them to adopt policies overturning traditional social relations or enlist revolutionary movements inspired by modern ideologies to replace traditional leaders with more accommodating native "progressives". Alternately, as the examples of the Americas, Africa, India, Southeast Asia and Australia all testify, direct imperial rule can dispense with indigenous sovereignty in either of two basic ways: A civil society's own underemployed can be settled on conquered territory in colonies exporting the metropolitan society while banishing surviving traditional communities to unwanted enclaves still subject to metropolitan rule; or indigenous leadership can be replaced with a colonial administration directly governing the conquered society in accord with the conquerors' interests.

Any of these forms of imperial domination can serve the economic needs that push civil society beyond itself. Yet in so doing, each form gives rise to something distinctly different from the modern society that autonomously arises from within a pre-modern setting. And in each case, the new predicament challenges modernity's normative tendencies to globalization with all too familiar legitimation problems.

Settler colonies present two normative challenges. Although the settlers may reproduce a civil society in which property rights, moral accountability, family freedom, economic opportunity, and legal equality are all upheld, so long as they remain subject to the rule of the metropole, their right to self-government is violated. That right counts, of course, only so long as political freedom is regarded as an end in itself (as Hegel's *Philosophy of Right* maintains) rather than as a means to guaranteeing the pre-political civil rights of individuals (as social contract theory would have us believe). In the latter case, benevolent and appreciated colonial rule can satisfy the liberal criteria of protecting person and property with the consent of the governed. If, however, the settlers are to enjoy the full breadth of modernity's embrace of freedom, which liberalism, in contrast to the Hegelian alternative, is incapable of completely legitimating, either of two options must be followed: the colony must be transformed into a province of the metropolitan nation, with the same political participation as any other, or the colony must become an independent democracy. Any other half way house, such as Puerto Rico's "commonwealth" status, violates modernity's norms for political right. Of course, because democracy is an ethical community that sustains itself only insofar as its members participate in self-government, the alternate options of integration into the metropole or independence can only provide a real solution if the settler population predominantly embraces one or the other.

A problem still remains even if national integration or independence gives a willing populace political freedom, at least so long as the indigenous population once banished to reservations has either chosen not or not been permitted to assimilate into the surrounding colonial society. If reservation life simply comprised an ethnic enclave in which the modern rights of property, morality, and family, civil and political freedom were exercised as in any other neighborhood or district, independence or integration would already resolve all abiding normative tensions. If instead the indigenous community follows traditional arrangements at odds with right, and/or remains governed by an external authority (e.g. a Bureau of Indian Affairs), a consistent realization of modern principles of right becomes problematic. Any external authority that lords over indigenous affairs deprives its subjects of the freedoms it usurps; yet that external rule may well reflect the unwillingness of the reservation population to give up their traditional community and assimilate themselves into the surrounding civil society and democracy that has been imposed upon them. Of course, the encompassing modern society must, to be consistent with its own principles of right, eliminate any barriers to such assimilation. Can, however, the encompassing modern society consistently retain rule over reservations in which tradition trumps right or consistently give complete independence to traditional communities in which modern freedoms are ignored, albeit with the consent of the members?

The latter question dovetails with the central dilemma arising when an autonomously emergent modern regime forces commercial concessions from a pre-modern nation or advances beyond such intrusions to conquer the resisting nation and impose direct colonial rule. In both of these cases, the modern intruder might regard itself as the bearer of the principles of civil society if it moved beyond the rape and plunder of the indigenous society (of which Marx accuses the British in the early expansion of the East India Company)[20] to the introduction and enforcement of universal property relations (i.e. including outlawing slavery, serfdom, and other forms of bondage), the recognition of moral autonomy (i.e. enforcing a separation of religion and state and religious toleration), the transformation of family relations in accord with principles of equality and co-determination, the enactment of freedom of occupation and public welfare guarantees, and the institution of a legal code and court system in which due process is respected. If these measures are not undertaken, the colonial power can hardly justify its usurpations other than on grounds of a national interest that turns a blind eye to the rights of others. Yet, even if the colonial power moves beyond plunder to forced modernization, the very externality of its intervention raises problems.

These problems are two-fold. First, even if it be granted that the rights of civil society (including property, moral and household right) are unconditionally valid, their very reality as modes of self-determination depends upon a preponderance of voluntary conformity by entitled individuals. This prevailing rectitude need not involve ideological acceptance, personal happiness, or any other specific psychological attitude. It does, however, require that individuals interact more in recognition and respect than in disregard of the appropriate rights of one another. Consequently, if the external imposition of civil society is to succeed, it must somehow achieve that degree of internal recognition on which its freedoms depend. This, of course, presumes, that the individuals involved are themselves acknowledged to be competent to exercise civil rights.

Can the latter acknowledgment be squared, however, with the deprivation of self-government that colonial rule incorporates? Admittedly, the pre-modern nation can hardly be said to afford its members self-rule if traditional governance is at odds with democracy. Yet colonialism's violation of the formal independence of the colonized equally fails to replace traditional tyranny with political autonomy. Arendt has observed that colonial regimes characteristically exercise an external bureaucratic management of their colonial subjects and succeed in doing so only by drawing a racial divide between the indigenous peoples and their metropolitan masters.[21] If the colonial rulers cannot ascribe racial inferiority to their subjects, how can they justify a domination that otherwise violates modern political right? Yet, as much as racism in colonial practice may pave the way for the introduction of racism into the internal political affairs of modern apartheid and fascist regimes, racism can hardly be consistently upheld by any regime that legitimates itself under the banner of modernization and development.

John Stuart Mill argued for a less inconsistent justification of colonialism: namely, that the subject peoples were not racially incapable of governing themselves but that they first had to undergo a tutelage weaning them from their oppressive cultural traditions before they would be able to participate in the institutions of freedom. Colonialism was therefore a temporary way station, superintending an external modernization that would culminate in independence once the formative process was over.[22] Hegel anticipates this view by arguing in paragraphs 350 and 351 of *Philosophy of Right* that the same unconditioned normativity of freedom that gives heroes the right to found states gives more civilized (e.g. more modern, free) states the right to regard the mores of traditional communities as inferior and to treat their independence as "merely formal",[23] implicitly sanctioning an external modernization through imperialism. If Hegel's basic insight be granted that political freedom cannot operate without the prior formation of a civil society, including universal property rights, recognition of moral autonomy, the triumph of freedom and equality among spouses in the household, the achievement of equal economic opportunity, and the institution of civil law, Mill's and his converging arguments would have some merit, provided two conditions could be met. First, colonial rule would have to be capable of externally fostering these developments without succumbing to a self-serving exploitation of the colony's wealth, and secondly, colonialism would then have to eliminate itself as soon as possible in a transition to independent post-colonial regimes.

To object that such a transformation violates the political freedom of the colonial people has a certain formality if the only alternative were a process whereby a pre-modern regime exercises its own non-democratic rule to modernize its society in preparation for metamorphosing into constitutional self-government. Without national independence, a subject community cannot possibly enjoy the political freedom that provides the capstone to modernity. Yet national independence alone may neither involve self-government nor a free society or household. External and indigenous rulers would both have to eliminate their respective forms of political domination if political emancipation is to succeed. The modernizing initiatives of an indigenous ruler would have no automatic privilege unless only an internal modernization can engender the level of voluntary cooperation required for the operation of the institutions of freedom or that external modernization can never resist the temptations of colonial exploitations. Although the situations are somewhat distinct, the examples of post-war Japan and West Germany suggest that external rulers can expeditiously eliminate the remnants of pre- and post-modern tyranny and reconstruct modern family, social and political institutions of freedom with widespread popular support.

Be this as it may, the genesis of the institutions of freedom can never conform to their actuality, because the volition founding constitutional democracy can never be an exercise of constitutional self-government. For this reason, neither route to independence can strictly accord with right any more than can the primitive accumulation helping form the first civil society. Since what is at stake in each case is realizing a good that is not yet at hand, the situation is one of moral accountability rather than of ethical responsibility, where one is bound by duties already realized in the institution to which one belongs. The prescribed task may consist clearly enough in contributing to the rise of universal property right, moral autonomy, household emancipation, civil society, and democratic self-government. Yet because the rights and duties contained in this result cannot be fully applicable until it is attained, their standard provides no unequivocal guidelines for what means to their implementation are genuinely necessary, let alone permissible.

Can the same predicament be said to apply to the post-colonial nation that a consistently modern colonizer must finally decolonize?

Principles of the Post-Colonial Predicament

In distinction from independent settler colonies, a post-colonial nation carries the burden of two decisive features: On the one hand, it contains the vestiges of pre-modern traditions, not isolated in reservations, but spread throughout society under the scaffold of a modern constitution mandating in varying degrees the institution of property, moral, family, social and political self-determination. On the other hand, the post-colonial nation's emergent civil society bears the burdens of competing economically, culturally, and militarily with established modern states. These "developed" nations enjoy the benefit of both a longer, more matured formation of capital and escaping the internal handicaps posed by the persistence of pre-modern traditions as well as the deformations created by the exploitation of colonial masters.

The resulting discrepancy in economic, cultural, and military influence between developed and "underdeveloped" post-colonial nations has led some post-colonial theorists to regard the coming of independence to be a purely formal transition, where colonial rule is supplanted by a neo-colonial domination changing the national identity of the local administrators but otherwise leaving the post-colonial masses in the same plight suffering with decimated traditions, chronic pauperization, and overwhelmed civil institutions struggling with corruption and bankruptcy. If Marx were right that capital dominates civil society and that civil society subordinates the state, independence would mean little more than this empty formality. The preponderant advantage of foreign capital would allow it to still control the decolonized society and utilize its newly independent government as no less a lackey for foreign exploitation as the defunct colonial administration.

The situation is very different, however, if Hegel is correct in 1) recognizing the ability of social interest groups, civil law and public welfare agencies to uphold the economic opportunity of all, and 2) recognizing the state's preeminence over civil society, not only in legislating and enforcing the laws that uphold pre-political freedom, but in regulating civil society to prevent social inequality from translating itself into political privilege and in thwarting foreign economic interests from dominating domestic society and politics. If the state can protect itself from domestic and foreign social domination, while insuring that social oppression is restricted and democratic freedom maintained, post-colonial political independence can have a double significance. First, as a harbinger of political self-determination, it represents the culminating phase of modernization, adding the final structure of freedom that pre-modern community has lacked: constitutional democracy. Second, it offers the post-colonial nation all the opportunities that political intervention provides for restricting the influence of foreign and domestic interests upon the development of its own society and political life.[24]

The history of colonial liberation reflects both of these gains. Although independence movements have followed various ideologies, they have generally embraced modern ideals of social and political justice, diverging primarily by giving different interpretations of what institutions are required to realize freedom. Whether inspired by Jefferson or Marx, the makers of independence have made themselves vehicles of modernization, bringing modern political relations (and their deformities) to their nation, rather than resuscitating traditional rulers and their communities. Indeed, the very appeal to nationalism has involved the introduction of a very non-traditional community uniting individuals in terms not of pre-modern hierarchies, but the equality of citizenship in a nation of one's own.

Further, each independent nation has attempted to use its own political regulation over society and commerce to orient its development away from the self-serving policies of past colonial masters. This may not be an easy task, but that it can be entertained at all suggests that the supremacy of state over civil society is not just a Hegelian dream.

Nevertheless, the future of post-colonial development is no more predetermined than the fate of freedom in the nations that first autonomously modernized. Fundamentalist and tribal reaction may succeed in tearing down the incomplete constructions of civil society and political democracy, just as a resurgent fascism may threaten the "developed" nations that fail to fulfill the family, social and political liberation that modernity has put before us. Whatever the future brings, the challenges of autonomous and external modernization will surface anew wherever the timeless normativity of freedom is rediscovered.

*****

The emancipation of reason and conduct from the hold of foundations represent only two sides of the project of modernity. Equally significant is the cultural arena in which the justification of aesthetic worth poses analogous problems. The next three chapters investigate how autonomy plays a central role in the features generic to fine art, in the differentiation of artistic styles, and in the unique way in which an individual art realizes the unity of meaning and configuration basic to aesthetic phenomena. The fitting place to begin is with an examination of the theoretical challenge posed by the individuality so crucial to the manifestation, creation, and appreciation of beauty.

Notes

1 See Alasdair MacIntyre, *After Virtue* (Notre Dame, IN: University of Notre Dame Press, 1981), 49 ff.

2 William Maker, *Philosophy without Foundations: Rethinking Hegel* (Albany, NY: State University of New York Press, 1994), 21–45.

3 To the extent that Hegel's *Philosophy of Right* identifies normative validity with self-determination and employs a putatively self-determining conceptual development to specify the institutions of right, that work's anatomy of property right, moral autonomy, a free family, civil society, and constitutional self-government provides a recipe for modernity on its own distinctive anti-foundational terms.

4 This is why Hegel is correct to remark that a constitution cannot be produced through an act of making, as Napoleon learned in trying to impose a modern constitution upon Spain. A constitution can only become more than a piece of paper if the non-political prerequisites of political freedom are already at hand, and precisely because non-political spheres are not directly administered by the state, they cannot be brought into being by political action alone.

5 G.W.F. Hegel, *Elements of the Philosophy of Right*, trans. By H.B. Nisbet (Cambridge: Cambridge University Press, 1991), paragraphs 341–360, pp. 372–380.

6 G.W.F. Hegel, *The Philosophy of History*, trans. by J. Sibree (New York: Dover, 1956), 9.

7 For if modern community inherently possesses a globalizing dynamic, any pre-modern community is liable to suffer that external impact, irrespective of its own internal metamorphic potential.

8 For this reason, recent United States' legislation depriving resident aliens of welfare benefits is contrary to the rights of civil society.

9 Hegel, *Elements of the Philosophy of Right*, paragraphs 245, 246, 248, pp. 267–268, 269.

10 Rosa Luxemburg, *The Accumulation of Capital*, trans. by Agnes Schwarzschild (New York: Monthly Review Press, 1968), 329 ff.

11 Hegel spells out the basic logic of these developments in paragraphs 190–195 of his analysis of the "System of Needs" in the *Philosophy of Right*.

12 Immanuel Kant, *The Metaphysics of Morals*, trans. by Mary Gregor (New York: Cambridge University Press, 1991), 191.

13 Kant, *The Metaphysics of Morals*, paragraphs 58 and 62, pp. 154–5 and 158–9.
14 Although Hegel undercuts the freedom of spouses by admitting a traditional hierarchy between husband and wife, he recognizes that the ethical community of the family is particular in scope. See Hegel, *Philosophy of Right*, paragraphs 161–172.
15 See the remarks to paragraphs 273 and 274, and the addition to paragraph 274 of Hegel's *Elements of the Philosophy of Right*, pp. 311–313.
16 See the addition to paragraph 274 of Hegel's *Elements of the Philosophy of Right*, 313.
17 Hegel, *Elements of the Philosophy of Right*, paragraphs 190–192, 241–245, pp. 228–229, 265–267.
18 In arguing for the inevitable turn to pre-modern markets and the inevitable collapse of capitalism when they have all been assimilated within the relations of capital, Rosa Luxemburg ignores the internal resources that public works, credit expansion, and deficit spending provide for capital accumulation. This is due to her reliance upon Marx' labor theory of value and the derivative reproduction schemes of capital developed by Marx in Volume II of *Capital*, both of which preclude these and other facilitators of economic growth.
19 Karl Marx, *Capital – Volume I*, trans. by Samuel Moore and Edward Aveling (New York: International Publishers, 1967), 713 ff.
20 Karl Marx, *Karl Marx on Colonialism and Modernization*, edited by Shlomo Avineri (New York: Anchor Books, 1968), 77–82, 84, 99.
21 Hannah Arendt, *The Origins of Totalitarianism* (New York: Harcourt Brace Jovanovich, 1973), 185 ff.
22 See John Stuart Mill, *Considerations on Representative Government*, Chapter XVIII, in John Stuart Mill, *Three Essays* (Oxford: Oxford University Press, 1987), 408–423.
23 Hegel, *Elements of the Philosophy of Right*, 376.
24 The richest and most competitive nations may promote a free trade ideology to roll back all such political intervention, but every independent state retains the option of regulating its own economy and its relation to foreign capital. Uncompensated nationalizations, tariff and import quotas, limits on foreign ownership of enterprises, flaunting patent and copyright agreements, and other subsidies for domestic commerce all have certain costs, but political independence alone makes it possible to wager these risks.

III
Beauty

Fifteen

The Individuality of Art and the Collapse of Metaphysical Aesthetics

The Rationale for Metaphysical Aesthetics

Aesthetics arises when dissatisfaction with our ordinary acquaintance with particular works of beauty drives reflection to ask, what is art? This submission to wonder involves not merely an urge to advance from uncertain opinion to knowledge about the particulars at hand, but a deeper compulsion to conceive art in its universality, determining what it is that makes something a work of art in distinction from any other kind of thing. Such a striving to determine art per se is not inherently metaphysical, even though metaphysical thought always begins by asking, what is? The manner in which this question is tackled is rather what differentiates the fundamental modes of aesthetics.

The metaphysical approach is distinguished by always answering the question of what a topic is by turning directly to given reality as its standard of truth. This makes perfect sense, provided one assumes that the universal character being sought is an intelligible nature, existing independently of our thought, but inherently conceivable by a passive exercise of contemplative reason. Such an assumption itself appears natural, for if some subject matter has no independently given intelligible nature, it is hard to see how inquiry into what it is can be rescued from the strife-torn relativism of conventional or purely personal identifications.

Yet applied to art, metaphysical reasoning might appear of little use. For whereas metaphysical thought seeks the measure of what is in what exists by nature, as something to be found in passive reception of intelligible form, art is expressly artificial, issuing from the conscious activity of individuals, the very source of everything arbitrary, perspectival and conventional. As it turns out, however, the dimension of art by which it is an artifact is perfectly suited to metaphysical construal.

To be an artifact is to be a product of craft, of purposive formative action that imposes an antecedently conceived form upon a given material.[1] The formative activity of craft creates neither the form it imposes nor the material on which it operates. Both are independent givens. Moreover, the form imposed upon the material, being conceivable apart from its realization, is something general in character. As such, it can inform many different materials so as to result in a plurality of artifacts of the same kind. What the artifact is is thus identifiable in virtue of the form it embodies, which determines it as one of a certain type. The material in which its form is realized may individuate it from other exemplars of the same form, but in no way contribute to its perfection, which depends instead upon the presence of its form.

Consequently, if one confronts an artifact and inquires into what it is, surveying its sensible qualities will be of no aid unless it leads one to think the form providing the purpose of the craft activity producing it. This is why an archeologist examining presumed artifacts can only understand what they are by discovering their function, which requires thinking the form they embody in common with any other artifacts of that type.

What makes the appeal to the paradigm of the artifact a characteristically metaphysical route is that it renders reality, be it of nature or art, something whose knowable character is rooted in antecedently given intelligible form. Although the artifact conception treats existence as if it were the product of artisans divine or mortal, the measure of reality remains something to be discovered, rather than constructed.

Employing the paradigm of the artifact to art has an immediate plausibility. Not only does it allow the produced work to retain a rational nature, susceptible of philosophical investigation, but it reflects a dimension of art that is ubiquitously acknowledged. Whatever else art may be, it always does comprise an artifact arising from an exercise of craft. Works of art are not simply natural things. They either issue as products of some sort of making, which, as in the case of "found" art, may even be so formal as merely taking a natural object and presenting it for display, or consist, as in the case of performance art, in an activity that is itself the work of art. Moreover, the artist, however original, cannot create *ex nihilo*, but must always work upon a given material, be it stone, canvas, film, the artist's own body, or language. For this reason, every artist must have a technique for mastering the given material and imprinting it with artistic form.

Nonetheless, that the work of art is unavoidably an artifact issuing from an exercise of technique by no means signifies that the nature of the art work can be reduced to this one dimension. The metaphysical approach characteristically attempts to conceive the *whole* of art in terms of the categories appropriate to artifacts and their production and evaluation. The success or failure of metaphysical aesthetics will thus largely rest on whether such categories are sufficient for grasping what is unique to art.

Art as Artifact

What then, are the basic implications of conceiving art as artifact?

First, with the art work conceived to issue from an exercise of craft, where an antecedently conceived form is imposed upon a given material, what the work of art is is defined by this antecedent form. It stamps the art work with an identity residing in an intelligible essence. Since this form is given prior to its particular embodiment, the material on which it is impressed contributes nothing essential to that form, but entails only indifferent accidents of its realization. Accordingly, the meaning of the work of art lies wholly in the antecedent form, which can be understood and defined apart from the artistic product that is its embodiment.[2] Moreover, this meaning is something of a general type, inherently reproducible in another embodiment without detriment to its significance.

Artistic creation, for its part, is here rendered a formative activity, guided by the prior apprehension of the form to be embodied in the given material of the work of art. Hence, the artist should know beforehand what he or she will make. Further, since this theme is defined by a form inherently susceptible of multiple embodiments, it can be known and communicated to others apart from its particular realization. Accordingly, the act of the artist, which consists of imposing this form upon a given matter, is limited to the execution of an idea previously conceived by the intellect. Producing neither the idea nor the material of the art work, the artist's act is confined to informing preexisting material. Artistic production thereby comprises a skill that can be taught like any other. All that is required to become an artist is learning the technique of how to embody a preconceived idea in a given material.

As for art appreciation, it now revolves around comprehending the antecedent form that gives the art work its meaning. Since the artist produces the work by following the guide of this preconceived form, the artist should know the meaning of the work. Art criticism, be it engaged in by artist or audience, will consist in evaluating how well the given form has been embodied in the given material. The work of art thereby is always to be judged by reference to an independently given end, which is conceivable in distinction from the sensuous apprehension of the work itself. In other words, art criticism becomes a matter of interpretation, uncovering through a labor of translation how the art work embodies a content that has a separate realization elsewhere.[3]

From the outset, these implications of the metaphysical construal of art as artifact appear fraught with problems. On the one hand, they seemingly fail to characterize art with sufficient specificity. If art is defined as a product of artisan activity, how can any distinction be drawn between works worthy of presentation in a museum and tools produced for any mundane use? What might otherwise be distinguished as the fine and useful arts seem to be placed on a par, leaving artistic creativity indistinguishable from artisanal skill and rendering aesthetic judgment no different than technical evaluation of a tool's functionality.

On the other hand, the application of categories of technique to art seems to contradict some of the key features that are commonly acknowledged to be generic to art.[4]

First of all, the work of fine art, unlike the product of craft, has an individuality whose unique originality is not an indifferent accident but a crucial contribution to the work's aesthetic value. For this reason, a copy of an art work, which may well exhibit completely comparable technique, lacks the imaginative originality by which it could command comparable aesthetic worth.

Secondly, fine art is marked by a unity of meaning and configuration, where the meaning of a work cannot be separated from its sensuous form. This unity is tied to the individuality of fine art, which insures that the significance of a work is not something shared by others and thereby distinguishable from them all as the common meaning they instantiate. Were the art work a mere artifact, with a nature common to other embodiments of the same form, its meaning would not only be devoid of originality, but something inherently separable from its particular realizations. In this connection, all particulars of the art work are essential to its significance, unlike a tool, whose reproducible function need not be impeded by the alteration of many of its sensuous features.

Third, given the unity of meaning and configuration of the work of art, the artist does not, and indeed, cannot have a preconceived idea of what the work will be prior to the act of artistic creation. An idea that is not already wedded to a sensuous configuration is not yet an artistic one. Hence there is no blueprint or formula that the prospective artist need only embody to create a work of art. On the contrary, when the artist produces something reducible to a formula that can be apprehended apart from the configuration of the individual work of art, the product may be a good example of craft, but it is a sorry failure as an example of artistic creativity. What makes artistic creativity worthy of the name is that the essence of the work of art is not given prior to its production, but rather arises within its realization, as it must to be in unity with every particular of its configuration.

Fourthly, since the artist creates without a preconceived idea exhaustively dictating the aesthetic features of what will be created, the artist need not be aware of what the finished work artistically signifies or how it should be aesthetically interpreted. Nor is it any artistic defect that the artist, as often lamented, understands not what he or she creates. An artisan, by contrast, would be poorly skilled if the function of his or her product remained a mystery to its maker.

Fifth, although artistic technique can be taught like any other craft, artistic creativity is a matter of genius, whose unique originality defies transmission through teaching. To be teachable, artistic creativity would have to be governed by general rules or common patterns. Yet given the individuality and unity of meaning and configuration of fine art, the creative act that produces it cannot be reducible to any rule-governed skill.

Sixth, art criticism properly judges a work of art in regard to its significant originality and internal unity of meaning and configuration. As we shall see, this does not entail excluding all connections with our world, but it does preclude centering the evaluation of art upon its accord with an independently given standard, like the governing purpose of an artisan production.

If the project of metaphysical aesthetics is not to be abandoned, these distinctive aspects of art must somehow be captured without transcending the nature of artifacts and artisan production. To this there can be but one solution: art must be distinguished from artifacts lacking aesthetic character by differentia identifying it as a particular species of artifact.

What can this species be, this brand of artifact that defines art, whose artisanal production distinguishes artistic creation and whose technical evaluation can comprise aesthetic judgment?

Art as Imitative Artifact

Plato provides the plausible answer: that the required species is none other than the imitative artifact, whose production is an activity of imitative craft and whose appreciation revolves around judging the fidelity of the imitation.[5] *Mimesis* or imitation hereby becomes a central category of metaphysical aesthetics, introduced to redeem *techné* or craft as a suitable framework for determining what art is.

Imitation in its aesthetic employment is not confined to the copying of another work of art, be the latter an artifact issuing from mortal artisanship or the craft of a divine demiurge who imposes form upon the chaos of matter. It rather extends to representation in general, to the extent that the representative function of mimetic craft involves the imposition of an antecedently given form upon some given material, where that form resides in the perceived appearance of something existing in nature or in another given artifact. The work that results "represents" the original not as a sign standing for it, but rather as an artifact reflecting the appearance of the original.[6]

Since the "original" that is imitated need not be an actually existing thing, but may well be another representation, whose own embodiment of intelligible form is given in a separate product of imagination, the imitation of mimetic art can be considered, as Danto puts it, nonextensional.[7] However, whether imitative art is about something real or merely imaginary, the key factor is that it always represents an independently given theme in terms of the craft categories of embodied form.

If the application to art of categories of craft entails such further appeal to imitation, the subordination of art under the category of imitation involves its own imposition of the categories of craft. To define art as an imitative object, produced by an activity of mimetic image-making, is automatically an embrace of the craft framework. Because the production of an imitative image must proceed from a given form that is then imposed upon the given material of the particular art, the mimetic artist produces art as an artisan. This can be true even if the likeness making involves not only the imitation of real things, but an imaginative, fantastic image-making, as Plato adds in the *Sophist*.[8] Indeed, if imitative art is to appear to its audience as a copy rather than as the original, there must be something about its appearance that distinguishes it from what is given in reality, something that stamps it, broadly speaking, as a work of fiction. No matter how fictional it may be, the work of art can still be produced as a copy of an antecedent form, provided that the resulting image realizes a preconceivable idea, be it through the combination or transposition of forms given in nature or convention. Moreover, imitation can equally apply to works that express inward states of character of the artist or other individuals, rather than describe external phenomena. Here again, artistry exhibits a mimetic craft to the degree that the produced image mimics the given reality of the soul.

Yet even if the concept of imitative artifact can accommodate the fictional and expressive dimensions of art, it seems to render problematic the whole rationale of artistic production. Robbing the work of art of a unique, independent significance, mimetic theory casts the value of art in doubt.[9] For the moment one grants that the work of art be an imitation of a given original, the question arises, for what end? Why copy what already exists?

Could it be for pleasure? This seems a suspect reason on two accounts. First, would not the furnishing of the imitated object be a more worthy satisfaction of desire than providing a mere copy? Secondly, the art object is characteristically not consumed like ordinary objects of desire. Instead, it is left untouched, as something simply to be let appear, and if need be, to be protected from personal use by the guarding conventions of a public museum or performance hall. Moreover, as Plato points out in the *Laws*, pleasure serves as a legitimating standard only for what furnishes neither utility nor truth of likeness, but only accompanying charm and amusement.[10] Yet art as imitative artifact must in some respect be judged according to how well it copies its original.

If not for pleasure, is art, qua mimetic, then for the sake of knowledge? Here we must contend with Plato's classic critique of whatever educational pretensions imitative artistry might entertain. Since the imitative art work, as itself either a sensuous object or words conveying sensuous phenomena, copies other sensuous things as they appear to the artist, it is automatically inferior as a source of knowledge to what it imitates. Not only is it at one remove from the original, and thereby something different, but it results from an attempt to copy the original not as it truly is, but merely as the artist perceives its sensible appearance.[11] And if the artist imitates not nature, but creations of other artists, the work is at a triple remove from the truth, comprising an imperfect replica of an imperfect replica of an appearance. Hence, if artists did aim at knowledge and really knew both what they were imitating and the character of their imitative artistry, they would be interested in the realities themselves, rather than in making copies.[12] Makers of imitative images thus reveal through their persistent practice of their own craft that they know appearances only. Whereas the users of objects must have true knowledge of them and their function to employ them properly, and the makers of useful objects must have correct belief about them gained from the user who knows, imitators can imitate without knowing anything of the true character of the objects they copy.[13] Furthermore, since what the imitative artist produces is not merely conceptual and objective to thought, but something imagined and appealing to sense and feeling, it encourages those irrational elements of the soul that are at furthest remove from the intelligible forms by which artifacts are knowable.[14] Creating works with an inferior degree of truth and appealing to parts of the mind that are cognitively inferior, imitative artistry is thus doubly unsuited for serving the pursuit of knowledge.

Edification as the Rationale for Imitative Art

What then remains as a reason for pursuing mimetic craft? Although imitative art may be inferior as a vehicle of knowledge to observation or thought, the appeal of its sensuous replica to sense and feeling makes it a useful instrument for cultivating those not yet ready and willing to undertake rational study of the true and the good.[15] By employing the suggestive powers of art, imitative artists can bring the young and others to take displeasure at ugly depictions of the false and base and pleasure at beautiful portrayals of the true and right while they are still unable to apprehend the reason for their respective repulsion and admiration.[16] In this way, imitative art can help mold the character of individuals by cultivating the proper emotive responses sanctioned independently by reason.

Although Plato and Aristotle both emphasize this edifying role of art in relation to the education of the young, it hardly follows that art should be a matter for youth alone. Imitative art can have the same salutary effect upon mature citizens who are capable of using their reason to differentiate illusion and truth and vice and virtue because the well springs of good living do not lie in reason alone, but require the consonant contribution of will. Since the training of desires to produce dispositions to take pleasure and displeasure in the right objects already inclines the individual to choose the proper course of action, art's appeal to sense and feeling gives it a power to influence character in a way reason cannot.[17] This influence may be of paramount importance in youth when character formation reaches a certain fixed plateau, but it retains a value that makes art of use throughout life. Through this edifying function, imitative art thus wins a very significant reason for being, placing artistic pursuit at the very core of individual cultivation.[18]

One is tempted to object that characterizing art as an edifying imitative artifact drastically restricts what creations can count as genuinely artistic. Readily representational media such as the visual arts and literature appear eminently suited to play their edifying roles, but what of such seemingly non-representational "arts" as architecture or music? Music is perhaps the most extreme example, since "pure" music, devoid of lyrics or program, seems incapable of representing anything given externally. Yet, Plato and Aristotle both turn to music more than any other art when it comes to forming the character of the young. Musical education is most sovereign for them because melody, harmony and rhythm reputedly take strongest hold upon the soul and combine to form modes and rhythms representative of different characters.[19] Hence, even though music is devoid of the spatial externality allowing the visual arts and literary imagination to portray objective reality, the musical modulation of time can mimic the temporal flow of the inner life of individuals, in all its emotive variety. Similarly, architecture's frozen music of harmony and proportion can shape its masses of heavy material to provide an environment for action reflecting the character of those who will inhabit it.

Granted the wide range of media susceptible of edifying employment, once edification becomes recognized as the only legitimating purpose of imitative art, the subject matter of aesthetic imitation becomes restricted to what can be edifying. Broadly speaking, imitative art will properly portray what is ethical. Since the ethical consists of normatively significant action in which ethically relevant character is displayed, art, metaphysically construed, will imitate neither nature in general, nor human affairs in their totality, but, as Aristotle maintains in his *Poetics*, *praxis* – understood as the sphere of ethical conduct and its institutions.

Moreover, since the edifying function of imitative art is essentially a public concern, involving as it does the formation of the right sort of character of citizens, Plato is perfectly consistent in calling for public scrutiny and censoring of artistic production. Given how a correspondingly metaphysical vision of the good life can take different teleological versions,[20] one might quarrel with Plato's edict that the only poetry worth admitting to the good state are hymns to the gods and praises of famous men,[21] just as one could object to Zdhanov's and Mao's decree that art embrace a socialist realism serving the masses. Nevertheless, art mimetically construed is just as appropriate a matter of political regulation as public education.

If art is of value for its edifying imitation, then it further follows that art criticism must consider first, whether the subject to be represented has the proper ethical character, and secondly, whether it has been imitated sufficiently accurately and with the kind of skill required for the work of art to have the desired edifying effect. To this end, art criticism will focus upon a meaning that can be formulated apart from the work. The aesthetic value of the work will depend upon the ethical significance of this independently given meaning as well as upon an evaluation of how well the work embodies this meaning in a salutary way. Hence, to paraphrase

Plato, the competent art critic requires three types of knowledge: knowledge of what the art work imitates, knowledge of the ethical truth of this original, and knowledge of how well it has been executed.[22] Significantly, since the work should influence the sensibility and feeling of its audience in a salutary way, what counts is not simply an accuracy providing epistemic satisfaction, but rather the sort of depiction that will produce the appropriate cathartic effect upon the emotions. When Plato prescribes how art should truly imitate the Divine[23] and what stories and narrative modes are appropriate for educating the guardians of the state,[24] he tends to emphasize the importance of correct depiction, as if this merely involved mirroring the given nature of the subject matter. However, if the edifying role of art is duly acknowledged, imitation can no longer be taken quite so literally. As we shall see, this consideration has fateful ramifications for the metaphysics of art.

Nevertheless, so long as the edifying role of art relies on imitation, artistic creation should presumably be predicated upon the artist's prior apprehension of what is to be depicted in particular regard to its ethical significance, as well as upon the technical skill and knowledge of how to mimic it in a particular material. Then, however, artists should know the meaning of their works and be able to teach others to produce equally beautiful objects. Yet, as Plato himself admits, artists characteristically do not know beforehand what they will create, nor, unlike other artisans, can they afterwards always explain their work, let alone be most knowledgeable about its significance, nor can their artistry be taught like other crafts. How can these features of artistic creation then be squared with the imitative edification ascribed to art?

Plato offers a solution indicative of what can be done within the limits of the categories of edifying mimesis. Artistic activity can retain its character as imitative craft while allowing the artist to create in ignorance if the artist be understood to create in a divine trance, where, without consciously knowing what is being imitated, the artist still executes a predetermined idea, working as the possessed instrument of divine inspiration.[25] If the appeal to divine artifice offends, one could just as easily follow Kant and substitute Nature, having it work through the unknowing artist, embodying preconceived form in some given material.[26] Either way, the work of art remains an artifact, whose significance can neither be foretold nor best interpreted by the artist, and whose production can hardly be taught to others by one who creates in a blind frenzy.

Aristotle and the Self-Transcendence of Imitative Craft

In proceeding from craft to edifying imitation, metaphysical aesthetics seems to have met success, constructing a coherent account of the work of art, art criticism, and artistic production. Yet, on closer examination, the metaphysics of art is rent by a troubling tension that threatens to explode the conceptual framework it has erected. Aristotle's *Poetics* brings this internal rift to a head by grappling with the problems left unresolved in Plato's seminal development of metaphysical aesthetics.

Like his mentor, Aristotle accepts the basic metaphysical notion that fine art is an imitative craft whose value lies in its edifying consequences.[27] Aristotle accordingly explains aesthetic value in terms of a natural propensity to imitate and take pleasure in viewing imitations, where the pleasure received is rooted in the satisfaction of learning something ennobling.[28] Thus music, for example, supplies aesthetic enjoyment by imitating all the qualities of character in its rhythm and melody, fostering habits of feeling pleasure or pain at its mere representations that cultivate the power of forming right judgments and taking delight in good character and noble actions.[29] Further, Aristotle limits the subject matter of fine art to ethical conduct (*praxis*), where the characters of agents get revealed in action in respect of their virtue and vice.

What marks Aristotle's development of metaphysical aesthetics beyond its initial statement by Plato is Aristotle's recognition that imitative art can only fulfill its function if it transfigures, rather than merely mirrors the given content of its artistry.

On the one hand, Aristotle keenly recognizes the implications of how art's edifying effect upon its audience is not purely intellectual, but involves its sensuous impact upon emotion. In the paradigmatic case of drama, the effect is achieved through the catharsis of incidents arousing pity and fear.[30] To achieve such catharsis, drama must not simply mimic any ethical conduct, but carefully choose its plot so as to produce the requisite reaction. Generally, drama must depict a certain type of ethical conflict and its resolution, involving a clash of ethical principles rather than merely natural calamities.[31] Moreover, Aristotle maintains, it will not do to portray a good man passing from happiness to misery nor a bad man passing from misery to happiness, nor an extremely vicious man falling into misery, nor pity occasioned by undeserved misfortune, nor the fear of one like ourselves since none of these cases inspire the proper fear nor pity. An edifying plot must instead involve an individual who is not pre-eminently virtuous and whose misfortune is brought upon not by any depravity, but by some great error that reveals a moral purpose.[32] Then, the drama can move its audience to identify with the protagonist and emotively respond to the moral conflict portrayed.

Yet, even here, plot construction is not simply a matter of copying appropriate actions. To achieve its artistic ends, dramatic art, like any other, must alter what it represents to give it a distinctly aesthetic form. To obtain the properly edifying catharsis drama must incorporate elements of explicit dramatic convention making clear to the audience that what they are witnessing is not real action, nor a natural object, but rather an artistic representation with its own fictional arena. Imitative art may indeed mimic appropriate subject matter, but artistic mimicry must be markedly different from its original simply to be recognizable as an imitation.

This difference, however, is not limited to strutting on a stage with all the external machinery of dramatic illusion. It equally involves a transformation of the action portrayed, whereby it becomes distilled into specifically dramatic action. This involves more than conforming to the formal unities of place, time and persona, and restricting the length of the performance in accord with the limits of audience retention.[33] Rather, plot must be infused with a special unity where all that occurs is necessary, yet not predictable. To achieve this distinctly aesthetic necessity, plot cannot simply imitate individuals in all the contingent and irrelevant detail of their existence. Instead, characters must be depicted for the sake of the action, leaving aside all features that are not pertinent to it.[34] Further, incidents should not be copied in the order in which they actually occur, subject to all the chance occurrences and intervening events that have little to do with the ethical conflict at stake. Each incident should rather arise as the probable consequence of antecedent incidents, with each revelation of character rooted in this connection.[35] Otherwise, artistic representation wanders from its edifying function, losing itself in an aimless diffusion.

The unifying thread underlying these transfigurations of the content of imitative art is set in relief by Aristotle's telling comparison of poetry to philosophy and history. History, not art, describes what has happened in all its particular detail. Because historical narrative is committed to representing what has been in its immediate givenness, the particulars it portrays are immersed in disconnected series of events, full of accidental and meaningless aspects.[36] Consequently, the particular content of history is not necessarily expressive of anything of universal, substantial interest.

Philosophy, by contrast, conceives what is necessary and universal, leaving aside the particular as something incidental to the realization of the universal.

Art occupies an intermediary position, for what art describes, Aristotle maintains, is not what has happened, but a kind of thing that might happen, a particular of a certain type whose possibility is probable or necessary.[37] Granted that a universal statement concerns what will probably or necessarily transpire to a kind of thing, the particular represented by art is therefore conjoined with what is universal. Unlike philosophy, however, art does not dispense with the particular. The work of art constitutively involves a sensuous content, be it realized in sensuous material, such as the stuff of buildings, sculpture, painting or music, or in sensuous images produced in the imagination through literature. Accordingly, art affects its audience in an emotive and not merely intellectual way. Nevertheless, unlike history, art represents particulars that are wholly infused with universality. In contrast to the prosaic reality described in historical narrative, art portrays particulars that in all respects exhibit what is universal and necessary. For this reason, although the work of art will affect its audience in an emotive way, this affect will consist in not mere feeling, but in a catharsis involving an intelligent grasp of what is universal in the sensuous configuration before it.

Owing to this key unification of universal and particular, art cannot be imitative in the sense of simply mirroring reality as it immediately exists in all its contingency. Instead, the imitation specific to art mirrors in sensuous form what is universal and necessary. This requires more than copying what is given to sense. To achieve its unique marriage of universal and particular, art must employ the imagination to produce sensuous images in which each detail is imbued with abiding significance. Since art can only accomplish its edifying function by representing ethical affairs, art's imaginative unification of universal and particular entails portraying interactions of particular individuals that convey universal, necessary conflicts of moral significance. To this end, dramatic plot must submit to just those sorts of constraints that Aristotle propounds to insure that each incident arises as the necessary or probable consequence of antecedent action.[38] Taken in earnest, the requisite unity entails that plot exhibit a heightened meaningfulness, where all details contribute to the action's central problem without ever becoming predictable or formulaic. In this sense, the necessity of plot development is not mechanical in character, exhibiting a form that can equally order a different particular content. Then the particular content would be a matter of indifference to the significance of the work, destroying its unity of sensuous configuration and meaning. Instead, plot, and indeed, artistic form in general, must exhibit an autonomous development, combining necessity and freedom such that every stage is both unexpected yet unavoidable the moment it occurs.

What allows art to fulfill its edifying role is precisely this imaginative melding of the particulars of action with what is universal and necessary. By uniting universal and particular in sensuous configuration, artistic imagination is able to convey truth more directly than history, yet without rising to the pure abstractions of philosophical thought. On the other hand, by retaining the particularity of the sensuous image, which has direct influence upon emotion and desire, art can influence action more directly than mere philosophical argument.

Yet if artistic imitation must transfigure what it represents so as to achieve a unity of particular and universal, artistic creation must undergo its own transformation. Instead of falling within the confines of rote craft, artistry must now combine two capacities. On the one hand, the artist must comprehend what is universal and necessary in ethical life and no longer what is just the given form of existing particulars. On the other hand, the artist must possess a power of imagination to construct sensuous particulars in thoroughgoing unity with the universal.

By the same token, the art critic can no longer determine aesthetic value by judging how well the work of art mirrors what is immediately given. The art critic must instead judge whether the artist has captured in sensuous form what is universal and necessary in the reality of ethical life, creating particulars that convey the pure possibilities of their type without ever appearing to be ordered by an external, separable formula.

Aristotle leads the metaphysics of art to this threshold, where the inherent requirements of aesthetic inquiry begin to transgress the confines of the categories of imitative craft. On all sides, the reality of art has come to entail a unity of particular and universal that seems fundamentally incongruent with the framework of edifying imitative art. The root of the problem is that art seems to require a melding of sensuous configuration and universal significance where, on the one hand, the particular can no longer be incidental to the realization of the universal, and on the other hand, the universal is such that it cannot be expressed apart from the particular sensuous content of the work of art.

This new unity not only challenges the scheme of embodied form underlying craft production, but it equally eliminates the problem of seeking an instrumental justification for the practice of art. Conjoining universal and particular, the work of art takes on an independent value because what it conveys cannot be expressed through other means. If the work of art's essence were merely universal, comprising a form both antecedently conceivable and realizable in other embodiments, then it would be dispensable.

The characteristic uniqueness of the work of art exhibits precisely this interpenetration of particular and universal, where any change of detail would detract from the work's meaning. Although any object is individual, the work of art hereby has an *essential* individuality, where what is unique about it is fundamental to its value. By contrast, a mere artifact can be variously altered without impugning its perfection since the reproducible form giving it its identity and function has no trouble surviving changes in the incidental details of its execution.

The categories of imitative craft are similarly challenged by how the work of art transfigures what it represents. What distinguishes artistic representation from any corresponding reality is not something on the order of the difference between the embodiment of form and the form itself. It rather involves an imaginative creation separating the world of aesthetic representation from the givenness of particular existence.

These aspects of the unity of particular and universal, individuality, and imaginative transfiguration all go together and all underlie a further feature to which metaphysical aesthetics has inadvertently lead: that the meaning of the work of art can no longer be conceived independently from the sensuous apprehension of it.[39] If art was imitative in the mechanical manner of craft production, then the work of art would be like a tool, becoming comprehensible only when its purpose is discovered, where this purpose is conceivable apart from the sensuous apprehension of any particular example fulfilling it. However, because art infuses every detail of its sensuous configuration with meaning, creates something unique, and transfigures, rather than rotely copies whatever it portrays, what a work of art signifies cannot be detached from its particular appearance.

To do justice to these features, features that metaphysical aesthetics is compelled to admit, the philosophy of art must break from the categories of craft by drawing a radical distinction between the useful, mechanical arts and the fine, creative arts.[40] The course of metaphysical aesthetics has itself revealed just how the teleological conception germane to craft production is applicable only to the useful, but not to the fine arts.[41] The conceptual framework of embodied form does have its place in construing the useful arts, where the artisan produces a functional object in virtue of antecedently apprehending the purpose that gives it its identity and form. In such craft, the entire formative activity of the artisan is governed by the understanding of this purpose, which is why the exercise of craft is eminently teachable and repeatable without detracting from its excellence. Further, the purpose, providing the form of the artifact, gives it an intelligible essence, definable in distinction from the accidental qualities that attend its realization in a particular material by the particular artisan. This intelligible essence, in which the meaning of the artifact resides, is particular in the sense of being distinguishable from the form of an object having a different function. However, it is not individual since every other artifact that is or may be produced according to the same purpose exhibits the identical form. Accordingly, when it comes to judging the useful artifact, the critic has an independently given, common standard by which to measure its success: the function dictating its form.

With fine art, by contrast, the categories of technique can only capture a very partial and subordinate aspect of the work, its creation and its criticism. What eludes the craft conception is the pervasive unity of universal and particular that resists the drawing of any distinction between an intelligible essence of the work of art and the sensuous accidents of its embodiment. If a putative artistic creation is still governed by the remnants of such a differentiation, it is a mark of aesthetic failure, signifying either an inability to rise above the trite, predictable formulae of handicraft or a surrender to allegory, where the meaning of the work is given independently of its sensuous configuration, which then forfeits its own value. A true work of fine art, however, exhibits a unique originality, where neither the meaning nor its embodiment can be detached. Owing to this unity of form and content, neither can be given prior to the artistic activity from which the work arises. Hence, the fine artist does not simply execute a preconceived plan in the manner of an artisan, but rather engages in a creative activity, for which no antecedent given can provide the measure of the finished work.

Nonetheless, the imaginative originality of the fine artist does not operate *ex nihilo*. The fine artist still employs a given material as an artistic medium, upon which the ideas of imagination must be imprinted. This is true even in literature, which might well be the most creative of arts.[42] Even something as artificial as language can never be a total creation of an author, as private language arguments underline. Fine art therefore always involves an element of technique that artists must master to be able to use a particular medium as a vehicle of creative imagination. Like any other technique, this can be taught and such skill, whether in the crafts of architectural drafting, stone carving, drawing perspective, color composition and paint mixture, harmony and counterpoint, or rhythm and meter, is what schools for the fine arts can reasonably aim to teach.

However, if technique is not totally supplanted by artistic creativity, it is subordinated as an instrument of aesthetic imagination. Technical mastery may be needed for an artist to give full vent to creativity, but brilliant technique cannot independently confer aesthetic value upon a creation, nor make its author a great artist. Although technique remains an indispensable vehicle, only through the original activity of creative imagination can the artist rise above being a virtuoso artisan. This creative facility, the genius of the artist, simply cannot be captured by the categories of craft because it consists not in embodying given form, but in creating the aesthetic form in the first place. The artist as creator eludes the metaphysical conception and calls for a revolution in aesthetics.

Finally, the demands of art criticism break the mold for evaluating edifying artifacts. If fine art is imbued with a creative originality in which what is universal and particular are inextricably joined, then art criticism must abandon the procedures suitable for judging functional products. Above all, the art critic must avoid evaluating the work of art in terms of any independently given criterion. No longer can art be judged according to its edifying effects or its success in conveying some antecedently given knowledge. Nor can the art work be deciphered as if it were an allegory, in search of a meaning whose essence can be detached from the accidents of its presentation. Instead, art criticism must grapple with a sense of aesthetic value that somehow resides in the unity of universal and particular specific to fine art.

This unity of aesthetic form is the fatal stumbling block at which metaphysical aesthetics arrives, exploding the limits of its own theoretical paradigm. What art as edifying artifact cannot suppress, yet cannot comprehend is the dimension of individuality, comprising the unity of universal and particular at the heart of the originality and freedom of artistic creation. In light of the breakdown of metaphysical aesthetics, the ultimate success of the philosophy of art will now depend upon developing the conceptual resources for conceiving individuality in the form in which it underlies the reality of art.

*****

Art's unity of meaning and configuration can take particular stylistic forms in function of the types of world view to which it gives expression and the types of configurations that best manifest such self-understandings. The foundation-free autonomy of modernity poses special challenges to artistic construal that the following chapter explores.

Notes

1 I am here freely drawing from the analysis of the categorial framework of ancient thought developed by Michael B. Foster in his *The Political Philosophies of Plato and Hegel* (Oxford: Oxford University Press, 1968).

2 This can be true both when the work is created as an imitation of an antecedently existent embodiment of a form, as in the Platonic scheme of art, or when the work figures as a representation of a prior form, where, as Nelson Goodman argues (see Nelson Goodman, *Languages of Art* {Indianapolis: Hackett, 1988}, p. 5), representation is not reducible to resemblance, and the work denotes what it represents without simply imitating it.

3 In this vein, Susan Sontag observes how the perseverance of the mimetic theory of art has condemned its followers to a perennial project of interpretation, for whose act of translation what always comes first is the search for the separable content of the work of art. See Susan Sontag, *Against Interpretation* (Doubleday: New York, 1990), pp. 4–5.

4 For a somewhat comparable list, see R.G. Collingwood, *The Principles of Art* (New York: Oxford University Press, 1958), pp. 20–26.

5 As Collingwood rightly emphasizes, Plato's appeal to imitative craft does not apply to the entire realm of art. (R.G. Collingwood, *The Principles of Art*, pp. 46–51.) It only extends to those arts that are representative, excluding, for instance, those types of non-representative poetry that Plato also acknowledges on several occasions in the *Republic* (392 D, 607 A). Whether Plato has any other conceptual tools at his disposal to determine such non-representative art is a question for speculation, since Plato's aesthetic analyses are almost entirely directed at works of representative, mimetic art. Here, in any event, Plato provides a classic statement of the metaphysical approach to aesthetics, even if he never advances it as an argument addressing art in general.

6 Although Collingwood maintains that representation must be distinguished from imitation in that art as representative relates to something given in nature whereas art as imitative relates to another artifact (R.G. Collingwood, *The Principles of Art*, p. 42), it makes more sense to regard imitation in Collingwood's restricted use as a particular type of representation, since artifacts and natural entities are equally susceptible of representation. On the other hand, since the imitative character of mimesis is commonly applied to its representative features in general, there is little reason not to allow the two terms to overlap. In any event, insofar as representation, broadly understood as the reimposition of given forms in given matter, remains bound to the categories of craft, which turn out to be unsuitable for determining aesthetic worth, Collingwood's claim that the aesthetic value of art cannot reside in its representative character (ibid., p. 43) retains its validity.

7 Arthur C. Danto, *The Transfiguration of the Commonplace: A Philosophy of Art* (Cambridge: Harvard University Press, 1981), p. 70.

8 Plato, *Sophist*, 264d–267d.

9 Or as Susan Sontag observes, the theory of mimesis sets art in need of a defense. See Susan Sontag, *On Interpretation*. p. 4.

10 Plato, *Laws*, Book II, 667b–669b.

11 Plato, *Republic*, Book X, 598.
12 Plato, *Republic*, Book X, 599.
13 Plato, *Republic*, Book X, 601.
14 Plato, *Republic*, Book X, 603, 605.
15 Plato, *Republic*, Book III, 401d.
16 Plato, *Republic*, Book III, 402a.
17 Schiller appeals on these same grounds to the edifying role of art as a remedy for overcoming the indifference of freedom and necessity, or of duty and inclination. See the Ninth Letter in Friedrich Schiller's *On the Aesthetic Education of Man in a Series of Letters*, trans. by Elizabeth M. Wilkinson and L.A. Willoughby (Oxford: Oxford University Press, 1986), pp. 59, 61.
18 Given that the life of the citizen provides the focal point of ethical conduct, it thus makes perfect sense to follow Allan Bloom in considering Aristotle's *Poetics* to figure as an appendix to his *Politics*. See Allan Bloom, *The Closing of the American Mind* (New York: Simon and Schuster, 1987), pp. 72–73.
19 Plato, *Republic*, Book III, 398c–400d, 401d; *Laws*, 667b–669a; Aristotle, *Politics*, 1339b.
20 I have described the basic logic of the metaphysical approach to ethics in Chapter 2 of my *Reason and Justice* (Albany: State University of New York Press, 1988).
21 Plato, *Republic*, Book X, 607.
22 Plato, *Laws*, 667b–669a.
23 Plato, *Republic*, Book II, 379a–382c.
24 Plato, *Republic*, Book III, 386a–398b.
25 Plato, *Ion*, 532h–536b.
26 Immanuel Kant, *Critique of Judgment*, trans. by Werner S. Pluhar (Indianapolis: Hackett, 1987), #46, Ak. 307, pp. 174–175.
27 Again, as with Plato, it must be borne in mind that Aristotle's advocacy of metaphysical aesthetics does not unambiguously extend to all of his remarks on art. As Collingwood points out, Aristotle, like his teacher, admits of a distinction between representative and non-representative art, allowing for the possibility of a type of art to which the categories of imitative craft are inapplicable. (See R.G. Collingwood, *The Principles of Art*, pp. 50–52.) Nevertheless, Aristotle's *Poetics*, at least in the fragmentary form in which it has survived, restricts its analysis almost entirely to the particular representative art that Aristotle finds exhibited in poetry and in dramatic poetry in particular. In this limited domain, Aristotle's discussion does provide a classic elaboration of the implications of the metaphysical approach in aesthetics.
28 Aristotle, *Poetics*, 1448b4.
29 Aristotle, *Politics*, 1339b.
30 Aristotle, *Poetics*, 1449a20.
31 Aristotle, *Poetics*, 1449a20.
32 Aristotle, *Poetics*, 1452b.
33 Aristotle, *Poetics*, 1450b25.
34 Aristotle, *Poetics*, 1449a20.
35 Aristotle, *Poetics*, 1452a.
36 Aristotle, *Poetics*, 1459a.
37 Aristotle, *Poetics*, 1451b.
38 Aristotle, *Poetics*, 1452a.
39 Michael B. Foster, *The Political Philosophies of Plato and Hegel* (Oxford: Oxford University Press, 1968), p. 201.
40 I am here following out the brief, but extremely important points that Michael Foster makes on art in the course of his masterful critique of ancient political thought. See Michael B. Foster, *The Political Philosophies of Plato and Hegel*, pp. 185–187, 201.

41 For this reason, Collingwood can urge abandoning employment of the term "fine art" for aesthetic productions because it retains a residual vestige of the technical theory of art. (See R.G. Collingwood, *The Principles of Art*, p. 36.) However, so long as "art" is still employed to describe certain artisan activities, "fine art" remains a helpful term provided it is used with cognizance of its incongruity with craft.

42 Foster, *The Political Philosophies of Plato and Hegel*, p. 186.

Sixteen

Romanticism and Modernity

Modernity as a Problem for Art

With the rise and global expansion of modernity, art has increasingly become a problem. Cast adrift from the fixed bearings of traditional shape and meaning while enduring the pressures of market necessity and public subsidy, art has confronted a dilemma internal to its own aspirations, calling into question the very significance of its enterprise. Through the crucibles of the Enlightenment, the Reformation, capitalism, the American and French Revolutions, and social democracy, a world has begun to come into being recognizing no other authority than the autonomy of reason and will. The rights to property, moral accountability, a free household, equal economic opportunity and self-government have become ever more acknowledged as universal principles revoking legitimacy from any particular factors given independently of self-determination, such as hereditary rank, race, gender, sexual orientation, or ethnicity. And through the widening struggles over erecting free households, civil societies and political democracies, the divide between the modern and the pre-modern has been drawn by a normative agenda in which the universality of free agency is to be realized in a self-sustaining system of institutions of freedom.

These very developments have thrown art into a crisis. For if art is to rise above entertainment and edification by giving an indispensibly revelatory sensuous shape to fundamental human truths, how is it to retain any significance when the self-understanding of humanity affirms the universal claims of autonomy and casts in aspersion the independent value of all external particulars? If every particular given is now incapable of providing an exemplary realization of freedom, how can art's sensuous configurations any longer lay hold of fundamental truths?

Of all philosophers, Hegel has most acutely confronted this problem. In his *Lectures on Aesthetics*, Hegel makes the dilemma of art in modern times the very core of his analysis of the romantic form of art. Under the rubric of romanticism, Hegel here systematically investigates what particular mode of artistic construal can be given the self-understanding generic to modernity. If we follow out the basic argument that Hegel develops, the ambiguous fate of art in modernity becomes apparent.

Hegel's Deduction of Romantic Art

Hegel's conceives the romantic form of art as the third and final mode of artistic construal, following upon what he calls the symbolic and classical artforms. These artforms are delineated on the basis of the possible relations between meaning and configuration in artistic beauty. Three basic options are conceivable: meaning and configuration can or cannot accord, and if they are incongruent, meaning may be either too abstract to fit the individuality of sensuous shape or too concrete to be limited to any particular appearance. This division becomes further qualified in light of the possible types of world view comprising the content of art and of the corresponding imaginative forms that can give them artistic expression.

In this way, the artform in which meaning is too abstract to match the individuality of sensuous configuration becomes more specifically defined as the symbolic mode of artistic construal in which the self-understanding of humanity invests essential value in an ideal that lacks the individuality of independent subjectivity, directing rational agency to contemplate its own essence in alien sub-rational forms of nature or in abstract powers that transcend given existence.[1] Similarly, the artform exhibiting congruence of meaning and shape, the classical, can only do so by expressing a world view that invests fundamental value in the outer manifestation of rational individuality and accordingly finds a perfectly adequate shape in the sensuous appearance of rational agency, shorn of physical blemishes indifferent to the expression of the soul, as well as of traces of inwardness at odds with the sphere of public worth.

What, then, is the corresponding meaning and shape with which artistic style can involve a content too concrete to be limited to any sensuous shape? It will have to entail a third option, one overturning the last holds of natural determination by investing ultimate value in a rational agency, which, in departure from classicism, now has its autonomy freed of attachment to any particular givens of sensible existence. All domains of self-determination can now have their rights acknowledged, including the spheres of inner faith and conscience. Not only does this confer equal entitlement upon all rational agents, irrespective of the sensible particulars of their appearance, such as race, gender, and birth, but it extends legitimacy to the pursuit of self-selected ends, to the degree that they can conform with those of others. Accordingly, not only do rights get extended to all competent independent rational agents, but these rights win an expanded content to include, alongside political and household entitlements, those of morality and civil society.

The ramifications for religious representation are equally familiar. On the one hand, to the extent that the divine is still honored, divinity will exhibit rational agency without being limited to a particular sensuous form. Hence, nature will be divested of any abiding divinity[2] and polytheism, with its attendant particularizations, will be supplanted by monotheism. Moreover, the divine will relate to individuals in respect of their autonomy, conferring absolute worth to them in their universal capacity as free agents, irrespective of their physical differences, and according them responsibility for their own integrity.[3] Consequently, the domain of the sacred can withdraw into the faith of individuals, leaving free an external secular sphere with a significance of its own. Indeed, for just this reason, the autonomy of individuals can be a focal point of concern without appeal to the divine.

Needless to say, these ethical and religious transformations comprise central agendas of modernity. This might lend support to the view that whatever artistic style is defined by shaping such a world view is not only a logical option of artistic form, but a historical stage in the development of art. However, even if modernity may be immersed in ongoing struggles to erect secular and sacred institutions conforming to the self-understanding of the third form of artistic construal, this does not guarantee that works in this style can only appear in the modern era or that art in our time can only be created in function of the prevailing world view. Artistic dissidents are no more inconceivable in any period than is the possibility of a responsive audience that bucks the tide of reigning convention. Nor does the historical emergence of such a world view preclude its eventual supplanting by any other type, together with corresponding artistic results.

Hegel, of course, identifies the above notion of rational agency with the romantic form of art and with the prosaic secular and religious aspects of modernity. As a consequence, he treats romanticism as the modern form of art per se[4] and suggests that its consummation of the logical possibilities of artistic construal also brings art to its historical culmination. This claim is presented in concert with questionable assertions that the conceptual ordering of symbolic, classical and romantic artforms follows a historical necessity and that the attainment of the romantic style brings art history to an end in that artistic creation will now be unable to break the hold of romanticism and produce in any further style.

Leaving aside these dubious qualifications, what is the new mode of romantic configuration that fits a world view for which no sensuous appearance can fully capture the worth of rational agency?

The Configuration of Content in Romantic Art

Although the new self-understanding of the significance of rational agency may devalue nature and the immediate existence of individuals, external appearance remains a factor, standing as the counterpart to the subjective inner domain of conscience and belief whose centrality can no longer be ignored. With the classical unity of sensible existence and rational autonomy disrupted, both sides take on a new character, altering the scope of appearances fit for artistic construal, as well as changing their relation to the meaning they express.

Whereas the classical ideal mandated that rational agency find its complete and valid objectivity in the external manifestation of activity, romanticism offers an image of self-hood that regards immersion in sensuous appearance as a shallow debasement of the internal dimension to which the individual must now turn to give life true value.[5] Yet this rejection of an embrace of corporeality hardly necessitates a rejection of conduct and of its concomitant engagement in the world. Whereas individuals must withdraw from immediate immersion in external appearance into their own inner life to find value, they equally face the challenge of intervening in outer reality to affirm and uphold the worth of their own subjectivity, as well as that of others. Thus, just as conscience may make inner motivation the key to moral integrity, acting to realize right intentions remains imperative. If all relation to outer appearance were instead severed, the possibilities of artistic construal would be undermined.

The abiding role of corporeal existence must therefore not be neglected in determining the romantic form. Hegel might appear guilty of such omission by identifying the principle of romantic art as inner subjectivity, according to which rational agency wins its true objectivity by turning away from external existence into the depths of its own feeling, belief and conscience.[6] Such a description could suggest that romanticism is an albatross, caught with a principle whose purely inward negation of sensuous immediacy is fit for thought alone.

Hegel, however, emphasizes that the absolute subjectivity of romanticism can only retain the element of self-imposed determination that autonomy involves if it gives itself a particular actuality, breaking into external existence, even while upholding its own inwardness as the focal point of its identity.[7] It is in virtue of this dual entry into and withdrawal from sensuous manifestation that inner subjectivity can provide the principle for a form of art.

Yet what are the ways in which such inner subjectivity can give itself an outer configuration susceptible of artistic construal? Somehow the corporeal appearance of rational agency must reveal the indispensability of an inwardness that cannot be adequately exhibited in any particular sensible existence. As Hegel suggests, the only way sensible appearance can play such a role is by presenting an activity in which rational agency supersedes the hold of external facticity upon itself and wins an autonomous existence.[8]

Given how the content of artistic beauty in general is determined, romantic art has three basic fields for configuring inner subjectivity: the domain of religious representation and the two aesthetically appropriable realms of ethical conduct, those of a world condition of revolutionary challenge, where the institutions of freedom are yet to be founded, and of a world condition of settled free ethical institutions. Hegel himself treats these options as the three phases of romantic art,[9] implying that they comprise a sequence both conceptual and historical. On this basis, Hegel offers a progression of stages beginning with romantic art addressing religious themes, followed next by romanticism depicting adventures of chivalry and courtly love, and concluding with a romantic art expressing what he calls the formal independence of character.[10] Such an ordering may well fit the history of Western art from medieval times through the Renaissance and into modernity, but any conceptual or historical necessity must rest on more than an empirical correspondence.

The Romantic Construal of Conduct

The problem of art and modernity comes most clearly into focus in considering how the content of secular conduct must be molded to fit the self-understanding expressed by romantic art. To affirm the principle of inner subjectivity, rational agents must engage in activity whose significance rests no longer simply upon its external ramifications, but essentially upon its relation to the autonomous inner aims and motives of individuals. Conversely, with nature disenchanted and naturally defined ethical community deprived of any exclusive hold upon normative conduct, the configuration of sensible existence is no longer of essential significance for rational agency. Because rational agents now find true value in the independent resolve of their own inner subjectivity, rather than in immersion in the outer appearances of conduct, the shape of the world in which they act and the physical appearance of their own selves are now left prey to contingency.[11] As a consequence, the content of conduct fit for romantic construal exhibits three basic features.

First, because rational agency here affirms its autonomy in overriding the hold of outer existence, the situations in which inner subjectivity manifests itself as the focal point of value can take on the most varied character. No physical environment and no dimension of the lives of individuals is either off limits or a privileged abode to the romantic artistic imagination. Whereas classical art linked aesthetic worth to an ideal of beauty ridding the appearance of humanity of its contingent blemishes, romantic art extends beauty to the ugly and ordinary, in which inner subjectivity is just as much at home.[12] Because the integrity of inner subjectivity does not hang upon the immediate appearance of existence, the configuration of the world is a matter of indifference until rational agency has put its heart into some outer affair, whose external visage even then cannot provide what is essential. Accordingly, action is no longer necessitated by any specific conflict or situation. It rather proceeds from the independent resolve of individuals, who set out on adventures simply in function of their conscience, passions and beliefs.[13] Consequently, romanticism's affirmation of the inner freedom of the individual leaves the world free to appear as its own given self.[14] In this respect, the content for romantic art is not generated by artists in the way in which creators in the symbolic and classical mold could conjure up themes nowhere encountered in experience. With the whole world the romantic artist's oyster, content can be completely prosaic in origin.[15]

Secondly, since autonomy is no longer subordinated to alien powers or to the realization of the externally given ends common to members of a naturally determined ethical community, agents can now reveal the gleam of their inward integrity in pursuit of the most varied self-selected particular aims. Protagonists can no longer be authors of laws and institutions, whose outer enactment is what gives their deeds significance. Right and wrong does not reside in the external object of action, but rather in inner motivation and personal judgment.[16]

Thirdly, since the freedom of agents is no longer limited to what their outer appearance can directly express, they now pursue their independent interests without fear of showing themselves replete with blemishes irrelevant to their inner autonomy.

These three features of the romantic construal of conduct get further qualified in terms of the two basic types of world condition that art transfigures: a "heroic" age, where the institutions of justice have yet to be instituted, and a settled world of ethical institutions, where freedom has won its self-sustaining order. Both world conditions are subject to romantic construal, even if romantic art is the preeminent modern artform and even if modernity might be considered the age where the principles of property rights, moral autonomy, the free family, civil society, and democratic government gain universal recognition and dominant reality with the global spread of regulated markets and constitutional self-government through revolution, imperialism and anti-colonialism. On the one hand, modernity may have made the institutionalization of self-determination the order of the day, but it has still been a time of struggle between competing forces where the freedoms of persons, moral subjects, family members, civilians and citizens remain subject to restrictions linked to factors such as gender, sexual orientation, caste, race, ethnicity, religion, class and economic inequality. Hence, even if romantic art were the exclusive prize of modernity, the modern age can still offer panoramas fitting each type of world condition. On the other hand, since inner subjectivity does not depend upon outer existence for all its content, it can conceivably affirm itself either in heroic conditions of revolutionary challenge or in the pedestrian setting of a constitutional state observing all the modes of freedom.

Romantic Options in a World of Revolutionary Challenge

A world condition of revolutionary challenge presents the romantic form with a variety of ways to shape conduct. A fitting one is the scheme of chivalry, which Hegel explores at length under the rubric of the second stage in the development of romantic art. Chivalry, as a mode of conduct, fits the romantic mold to the degree that it presents knight errants pursuing self-imposed aims of personal honor, faith, fidelity, and courage, where the course of events follows the contingencies of adventure, rather than the fateful conflicts of ethical communities, torn asunder by conflicting pathos rooted in different institutions. In such sagas, what counts are not the achievement of deeds reaffirming ends existing in bonds of actual communities, but rather the singular strivings of individual protagonists, running after a glory that attaches to the affirmation of their own heart and passion.[17] It matters not whether the scene be medieval Europe, the wild west, the high seas, savage jungles, arctic wastes, or extraterrestrial outposts. In each case, chivalry can proceed provided the adventure turns upon the testing of individual fortitude in self-imposed challenges.[18] What counts as a collision engendering action is not the objective reality of a perceived injury, but rather the internal feeling of being wronged, which can be elicited by any circumstance, especially when honor and love depend upon the recognition of others.[19] Since chivalry thereby proceeds from contingent situations of no inherent importance to achieve ends of a purely particular and subjective character, it tends to subvert its own high pretensions and set the stage for a comic treatment of its self-deceiving nobility.[20]

Chivalric possibilities are also possible in dramas of revolutionary upheaval, where individuals participate in founding institutions of freedom, or in dramas of reform, where protagonists struggle against particular shortcomings in existing institutions where oppression maintains a partial reign. Since chivalry can invest personal honor in punishing any perceived injury, the righting of public wrongs can readily be made a personal crusade. However, using public struggles as an occasion for testing personal honor runs the risk of trivializing the commitment to justice and thereby unveiling the formal vacuity of chivalry's absorption in being true to one's heart. Moreover, the success of revolution brings to an end the significance of chivalry to the degree that the founding of the institutions of freedom upon the ruins of overthrown tradition upholds the value of inner subjectivity through a self-sustaining objective structure no longer depending upon any singular exertion for its maintenance. By the same token, the success of reform either remains limited to single instances of limited and transient importance or else it introduces the institutional change that allows the upholding of right to be rescued from the contingent mercies of the kind of subjective engagement on which it is based.

What, then, can the romantic imagination do against a backdrop of post-revolutionary institutions, where an established political democracy, civil society, and free family institutions provide an existing order in which personal autonomy is ratified? Where, in such a world condition, can the affirmation of inner subjectivity in *individual* conduct any longer retain fundamental significance?

Romantic Options in a Post-Revolutionary World

One option for romanticism is to focus upon the periphery of the settled institutions of freedom in search of a persisting frontier where the unique appearance of individual actions still exhibits an essential value. This frontier can exist in a spatial or a temporal detachment from the prosaic reality of a constitutional state upholding all the rights and duties of individuals.

Spatially distinct frontiers can be imaginatively reached through voyages beyond the boundaries of prosaic "civilization" to either unsettled territories or to communities that lack institutions of freedom.

Temporal distancing provides two alternate possibilities. On the one hand, the scene of action may be set in the period when individuals make their entry into the institutions of freedom, passing through the time of youth and apprenticeship. Romantic art can still find a place for inner subjectivity in this gap between the unsituated, indeterminate possibilities of youth and the "iron cage" of institutional immersion and adjustment endemic to any real engagement with the world. However, once youth is traversed and individuals occupy their roles in standing institutions of freedom, their expression of inner subjectivity in their individual deeds takes second stage to the impersonal workings of civil society and constitutional democracy.[21]

On the other hand, artistic imagination can take its protagonists on a voyage in time, withdrawing them from their place in a modern world to different ages of the past or future in which the conditions of chivalrous adventure reappear. Needless to say, whatever adventures ensue take place in light of an abandoned world whose objective machinery of law and order belittles the significance of any single subjective striving.

This contrast becomes all the more poignant in the reverse situation, where, instead of transporting characters to spatial or temporal frontiers, romantic imagination presents the quixotic plight of anachronistic figures who affirm their inner subjectivity as if the world could still allow of chivalry.[22] In this case, the object of romantic art becomes the dilemma of attempting to uphold the centrality of a self-imposed honor in an era in which the rights of conscience and individual interest are already actualized by, on the one hand, a civil society allowing individuals to satisfy their self-selected needs in reciprocity with one another, and, on the other, by a state ruling its citizens through laws they have democratically imposed upon themselves, with due respect for freedom of belief and the importance of motive in assigning criminal guilt.[23]

Yet if spatial and temporal distancings are not relied upon to create new frontiers, what opportunities remain for romantic construal in face of functioning institutions of freedom? At most the pursuit of personal honor, love, and fidelity involves purely particular aims and passions, in contrast to the universal reach of civil and political institutions, whose self-sustained constitutional operation no longer rests on any single agent or deed. Even if characters persist in holding fast to youthful strivings of world-reform, the recalcitrant apprentice will either remain a quixotic figure, whose modern knight-errantry comes too late, or a mundane activist who embraces the modalities of civil and political reform in which impersonal organizations and institutions again take center stage.[24]

The Formal Independence of Romantic Character

A configuration of conduct in a post-revolutionary setting more suitable for romanticism would be one in which rational agency champions the particular content of its inner subjectivity, irrespective of its situation within a world of existing institutions of freedom. Inner subjectivity would then affirm itself in such a way that it is free of bondage to any particular external existence and thereby at liberty to utilize any embodiment with equal indifference. Inner subjectivity then manifests an autonomy detached from any existing mooring, yet restricted to a formal independence to the extent that no shape it gives itself is adequate to its freedom. Its character accordingly comprises a personality defined purely subjectively through its own arbitrary amalgam of particular emotions and aims. These, as subjective, are independent of any externally given content of essential substance, such as would inform the pathos[25] of an exemplary protagonist.[26] Individuals of such character therefore have two basic options for affirming their governing inwardness: either they seek to realize their capricious subjective aims with the whole force of their personality or they turn inward to preoccupy themselves with the inner depths of their conscience, emotions and beliefs.[27]

In the former case, which Hegel duly analyses under the rubric of the formal firmness of character,[28] individuals act upon the world on the basis of their own contingent particularity, which they treat as independently being of highest value. However, precisely because this prized particular element of each personality is not inherently linked to any existing institutions, divinity or other external factor of universal significance, action in its pursuit is prone to encounter obstacles whose overcoming can only be a matter of accident. Hence, the portrayal of this one-sided, subjective resoluteness will entail a self-engendered adventure replete with conflicts. Any hope of an abiding reconciliation between the world and individual's strivings to assert inner subjectivity is vain since the content of personal aims and motives is just as capricious as the fickle contingencies with which the agent must cope.[29]

If, alternately, the individual be configured as one who withdraws into inner life without seeking its outward expression, the occupation of inwardness with itself can only warrant primacy if it possesses a matchless profundity.[30] Yet because such individuals remain situated within a world to which their inner preoccupation provides no harmonious bridge, they are liable to fall prey to colliding external contingencies with which they must make due.[31] Thus, this form of character has a twofold outer manifestation susceptible of artistic construal: first, the expression of its self-engrossed inwardness and secondly, the collisions it experiences in its relation to an indifferent world.

Focused inward or outward, the autonomy of such individuals remains a negative freedom, devoid of any positive external filling of inherent worth. The challenge facing romantic construal is not so much a matter of upholding the validity of this mode of freedom, but rather the question of just how significant any *individual* engagement in such moral self-determination can be against a backdrop of settled institutions of freedom. Since those institutions sustain themselves through lawful self-government, individual exemplary conduct stands relegated to an example of rectitude, rather than providing the rule for life. Although constitutional institutions of freedom cannot sustain themselves unless a certain level of rectitude predominates among the citizenry, the transgressions or obedience of one individual are no longer of essential importance.

This diminished significance of individual conduct for the fabric of the ethical world is reflected in how outer appearance is here set free in an independence possessing the same formality as that of the contents of inwardness. Although the configuration of art can take any shape whatsoever and still stand as an expression of inner subjectivity, this latitude is just as much a mark of indifference, exhibiting the leveling of all configurations into dispensable vehicles of aesthetic imagination. The course of action in which inwardness reveals itself thereby becomes a contingent adventure whose circumstances and complications are as devoid of inherent significance as they are unrelated to the aims pursued in their context.[32] Consequently, the resulting trials and tribulations of inner subjectivity are ready made for a comic treatment, setting in relief the contingency of inner and outer and the discrepancy between the absolute claims of inwardness and the capricious character of its content.[33]

Hegel characterizes this dual determination of meaning and shape as the formal independence of individual particularities, introducing it as the third stage in the development of romantic art,[34] taking as its arena a post-revolutionary world of settled institutions of freedom in which conduct can affirm inwardness either by acting upon the world with subjective resoluteness or retreating into a preoccupation with conscience, feeling and belief.

Realism, Irony, and Non-Objectivism: The End of Romantic Art

Although these may be the remaining options for romantically construing conduct in its own right in a post-revolutionary environment, romanticism can inform artistry's own expression in the work of art with the same configuration of rational agency. The same formal independence of character can be exhibited by depicting the contingent givenness of existence as an indifferent vessel for the expression of the artist's own inwardness or, conversely, by subjecting appearances to a comic manipulation, exhibiting their superfluity for the artist's inner subjectivity. With the appearance of individual conduct within a constitutional order of freedom no longer of essential importance to the affirmation of inner subjectivity, romantic art can express this very truth by embracing the given content of existence and celebrating how it, in all its indifferent detail, reflects the secured reality of inner subjectivity, or alternately, by subjecting the strivings of individuals and every aspect of reality to comic relief and subjective play. Hegel aptly suggests both as abiding options.[35]

The former option seems to put into practice the theory of mimesis, whose inability to grasp the individuality in beauty left it incapable of either justifying the value or conceiving the specificity of fine art. Hegel admits that romantic construal here reverts to imitation, but he strictly demarcates its depiction of given appearance from the edifying idealization of classicism and from the rote description of craft illustration and naturalism. In contrast to the classical appropriation of the given configuration of rational agency, romanticism appropriates the world of humanity, not in function of its inherent necessity, but in terms of its pedestrian contingency.[36] However, this portrait-like[37] embrace of the mundane is still distinct from the slavish homage to immediate appearance characterizing rote illustration and naturalism because romantic "imitation" makes an abiding attempt to infuse mundane particularity with the mark of inner subjectivity.

Such infusion can work in various ways. On the one hand, it can operate by injecting the most prosaic scene with a feeling reflecting the worth of inner subjectivity. This feeling may be manifest in the expression of rational figures who show themselves to possess an inner life accompanying the most mundane pursuits.[38] Or, the feeling may reflect the view of the artist, who takes configurations of no inherent human significance and transfigures them with a felt intensity manifesting the centrality of the heart of their artistic creator. Alternately, the appropriation of given appearances can exhibit nothing but the subjective mastery of the artist, whose absorption in reproducing the most evanescent moment or the most detailed panorama celebrates an independent creativity that can affirm itself in each and every content. Or, to similar effect, individual works and styles of other artists can be appropriated to set in relief how they get newly transfigured in the imaginative play of the artist. The "discovery" of "found art", impressionism, photorealism, the aping of familiar commodities, and the "post-modern" quotations of past styles, to cite a few familiar artistic strategies, all exemplify at least in part this type of romantic construal.

Since in all these cases what is at the crux of display is the inner life of the artist, rather than the content of the objects depicted, it is perfectly logical for romanticism to advance from exercising subjective mastery over the manner of representing prosaic things to gain control of the subject-matter as well.

A first step in affirming the rule of inner subjectivity over artistic theme involves imaginatively negating the putative independent significance of the outer world. Hegel duly characterizes this option as the free reign of irony, which debunks everything claiming external substance by rendering it a plaything of an unmasking artistic subjectivity.[39] The romantic art of irony above all presents the subordination of external contents and their related forms to the particular personality of the artist, whose conscience, passions and beliefs are given primacy.

Taken to its logical conclusion, this mastery over content can forego irony's dependence upon the themes of which it makes sport and employ completely non-objective images, whose only source of meaning is the subjectivity of the artistry they express and the subjectivity of the audience to whose imaginative play the work is relinquished.[40] With this extreme of abstract-expressionist, non-objective art,[41] the content of art becomes a pure appearance, setting artistry itself at center stage, producing a new objective world out of its own inwardness.[42]

In all these variations,[43] the contrasting poles of inner subjectivity and outer configuration exhibit a common formal independence, reflecting the underlying connection between the romantic flight to inwardness and the embrace of, on the one hand, the most dogged realism, and, on the other hand, the most extreme abstract-expressionism.[44] In these ways, of which Hegel only provides a partial account,[45] the third form of romantic construal exhausts romanticism's last remaining field for artistic transfiguration.[46]

The Production and Reception of Romantic Art

The production of romantic art involves a freedom of artistic creation exceeding the autonomy of symbolic and classical artistry just as the romantic vision imbues rational agency with a self-determination surpassing any conferred upon humanity by the symbolic and classical world views. Romanticism frees the artist from all external restraints and grants the artist the liberty to make the work of art nothing but the embodiment of his or her own creativity. This dual liberation comes to the fore with the third form of romanticism, where the content and form of the work of art are stripped of all independent prescriptions and made wholly responsive to the artist's personal choice of how to express his or her own subjectivity.[47] With every topic and every mode of treatment at the free disposal of the artist's inwardness, the romantic creator is preeminently *self*-determining,[48] engaging in an artform that has the unique character of perennially canceling its own specificity, subverting all given limits that might guide artistic creation. In this way, the romantic artist must confront more radically than any other the question of what are the proper boundaries of art, with respect to both the content of art and artistic form. Indeed, it might be more proper to say that, ultimately, the practice of romantic creation does nothing other than raise this issue ever anew, if always within a horizon that leaves art hostage to the liberty of subjective inwardness. For this reason, Danto is not entirely off the mark in suggesting that modern, that is, romantic art, converges with the philosophy of art.[49]

Accordingly, the reception of romantic art must grapple with the same fluidity of boundaries, the same arbitrariness of form and content that bedevils romantic creation. Whereas the audience for symbolic and classical art need not be plagued by doubt as to whether an object is a legitimate candidate for aesthetic appreciation, the audience for romantic art cannot avoid confronting the problem of judging how to identify art and the artistic construal of inwardness. The difficulties may be comparatively minor when encountering romantic arts of religious representation and chivalry, since these incorporate a settled content and configuration. When, however, the third form of romanticism is at hand, each act of reception may well involve a renewed judgment of where lie the limits of art.

Romanticism and the End of Art

Granted that symbolic, classical and romantic art exhaust the three basic ways in which meaning and configuration can be united in artistic beauty, the third stage of romanticism equally consummates the determination of the particular forms of art.

Hegel, however, adds two key additions to the sense of this consummation. He claims that this third romantic mode comprises the historically final development of art and that it not only comes last, but brings art to an end.[50]

This ushering in of art's end would be most easily plausible if all it signified is that the advent of this third mode brings to appearance all of the possible forms of artistic construal. In that case, although artists could continue to create, all further production of artworks could do no more than repeat modalities that have gone before. Hegel, however, is not just denying that art can invent new modes. He is arguing both that romantic construal must come after symbolic and classical styles and that the end of art achieved by the formal independence of individual particularities signifies the elimination of the very stature of art as a vehicle of fundamental self-understanding.

The former claim of historical finality is dubious for the same reasons that undermine Hegel's claims concerning the necessary historical ordering of symbolic and classical art.[51] The world view affirming inwardness may well arise through a rejection of the symbolic homage to an abstract absolute devoid of individuality or of the classical celebration of the immediate external existence of rational agency. Yet even if this were the case, it does not guarantee that romanticism need arise in that way nor that it emerge after the other modes of artistic construal have run their course. All could conceivably coexist in an age of competing cultures or at any time among dissident creators.

However, admitting that romanticism can conceptually exhaust the possibilities of artistic construal without necessarily enjoying historical finality does not rule out Hegel's further claim that the third variety of romantic style brings art to cancel its own constitutive significance.

Hegel's reasoning is simple enough: Because this mode construes the external circumstances of conduct to be just as contingent and willful as the aims and purposes of inner subjectivity,[52] the two sides of outer and inner relinquish the last trace of unity on which beauty depends. With rational agency withdrawing into itself and leaving the external world free to go its own way, what appearances get chosen to embody inner subjectivity has become as completely arbitrary as the content of inwardness, leaving meaning and configuration wholly external to one another.[53] The unity of artistic beauty, however, cannot subsist if shape and significance are rendered so capricious and indifferent to one another. Accordingly, Hegel concludes, romanticism's striving to configure the independence of subjectivity ends up revealing that art cannot succeed in conveying the full truth of rational agency. If humanity is to achieve genuine self-understanding, it must therefore turn away from art to some more adequate vehicle of comprehension.[54]

What is so special about the third form of romanticism that brings Hegel to announce the end of art?

In one respect, this third form seems merely to renew the opposition with which romantic art has grappled from the outset: the opposition between external appearances set free of any immediate unity with rational agency and an affirmation of inwardness which, though requiring expression, is inherently incongruent with every outer manifestation.[55] Hence, it is tempting to conclude that the formal independence of individual particulars actually retains a unity of meaning and shape since the freedom with which external existence can be employed conforms to the arbitrariness of the content of inner subjectivity.

Hegel, however, signals the emergence of a new predicament. With the elimination of all limits upon which particulars serve to express inner subjectivity, the two sides have fallen completely apart. Now, the more subjectivity has its role enhanced, the more external appearances are deprived of inherent meaning. Consequently, the choice of theme and manner of depiction becomes a matter of personal arbitrariness, subjecting every objective content and form to playful manipulation and distortion by the artist's subjective mastery.[56] With romantic artistry divesting every theme and every manner of treatment of any inherent worth or any inherent connection with its counterpart, art's unity of meaning and configuration would seem to have been brought to complete collapse.

No artistic solutions seem possible on two accounts. On the one hand, all the fundamental modes of artistic construal have already made their appearance. To this extent, no new options are available. On the other hand, the romantic artist cannot coherently revert to symbolic and classical construal since the conceptions of humanity generic to those artforms remain repudiated by the romantic world view. If there is to be any reconciliation, it will have to occur outside art in some more suitably inward vehicle, such as religious belief or speculative thought.[57]

Yet do these difficulties signal the collapse of art, as Hegel suggests, or do they rather comprise the perennial challenge facing artists in a modern world? Hegel himself steadfastly grants artistic status to the two forms of romanticism that he identifies as the dual aspects of the dissolution of art. He will not deny the ascription of aesthetic worth to creations that infuse contingent given appearances with the mark of inwardness nor to those that subject all considerations of content and form to the free play of the artist.[58] In each case, a modicum of aesthetic unity persists insofar as the very contingency of the content matches the subjective mastery in the manipulation of form. Not only are subject-matter and treatment equally capricious, but both thereby express the same affirmation of a subjective autonomy that is not essentially wedded to any external reality. This may well signify that the object represented by art has lost all absolute interest in itself.[59] However, if what counts instead is the inner subjectivity of the artist as it triumphs over the sensible appearance of its expression, this could just as well comprise art's greatest success in conveying the truth of rational agency.

Nevertheless, even if the "end" of romantic art retains aesthetic value, its liberation of art from the hold of any given form and content still offers only a truncated view of humanity. For if inner subjectivity can only be artistically construed in an otherworldly religious dimension, in the inherently contingent personal crusades of chivalry, or in the formal independence of character as it affirms inwardness in indifference to existing institutions of freedom, art fails to grasp the full truth of how inner subjectivity comprises only an element of rational agency, an element in need of incorporation within a self-sustaining constitutional democracy. It is that incorporation that diminishes the significance of the individual appearance of conduct and that thereby leaves the romantic style with no unique image that it cannot cast aside.

*****

Whereas modern art puts the aesthetic unity of meaning and configuration to its ultimate test, architecture presents an imposing challenge to every artistic style, given the utility of buildings and the abstraction of their heavy forms. The next chapter investigates how architecture can command aesthetic worth by shaping environments in expression of the values prescribed by different world views.

Notes

1 Hegel, *Lectures on Aesthetics*, trans. by T.M. Knox and W. Miller (Oxford: Oxford University Press, 1975), I, p. 517.
2 Hegel, *Lectures on Aesthetics*, I, p. 524.
3 Although Christianity may express this respect in the narrative of Christ, where the divine exists in an actual historical individual, Hegel's suggestion that religious representation must adopt such a narrative in order to uphold the significance of autonomous individual agency is far from evident.
4 However, Hegel does include within Romanticism all art of the Christian era, which, he maintains, from its outset begins the progressive working out of a free spirituality liberated from nature, whose full sacred and secular consequences will only be achieved in modern times.
5 Hegel, *Lectures on Aesthetics*, I, p. 518.
6 Hegel, *Lectures on Aesthetics*, I, p. 518.
7 Hegel, *Lectures on Aesthetics*, I, p. 519.
8 Hegel, *Lectures on Aesthetics*, I, p. 522.
9 Hegel, *Lectures on Aesthetics*, I, p. 528.
10 Hegel, *Lectures on Aesthetics*, I, p. 528.
11 Hegel, *Lectures on Aesthetics*, I, p. 524.
12 Hegel, *Lectures on Aesthetics*, I, pp. 526–527.
13 Hegel, *Lectures on Aesthetics*, I, p. 589.
14 Hegel, *Lectures on Aesthetics*, I, p. 527.
15 Hegel emphasizes how romantic art takes its content from the externally developed doctrine of Christian religion. (Hegel, *Lectures on Aesthetics*, I, p. 526.) Although this may be historically the case for European Medieval art in particular, it is hardly a necessity for romantic art in general. Not only is Christianity just one conceivable example of a religion fit for romantic construal, but romantic art is free to draw its material from purely secular affairs of the most wide variety.
16 Hegel, *Lectures on Aesthetics*, I, p. 590.
17 Hegel, *Lectures on Aesthetics*, I, p. 528.
18 As Hegel suggests, the Crusades provide a paradigmatic adventure of chivalry insofar as its participants banded together voluntarily as individuals in pursuit of an end whose debasement of something spiritual into a physical possession left no prize of any worth but the winning of personal honor. (See Hegel, *Lectures on Aesthetics*, I, pp. 587–588.)
19 Szondi, *Poetik und Geschichtsphilosophie I* (Frankfurt am Main: Suhrkamp, 1974), pp. 448–449.
20 Hegel, *Lectures on Aesthetics*, I, pp. 590–591.
21 Hegel, *Lectures on Aesthetics*, I, p. 593.
22 Hegel, *Lectures on Aesthetics*, I, p. 591.
23 For a discussion of how civil society and democratic government provide institutions of ethical community in conformity with moral subjectivity, see Richard Dien Winfield, *Freedom and Modernity* (Albany: State University of New York, 1991), pp. 70–71.
24 Hegel discusses such a possibility in his treatment of romantic fiction. See Hegel, *Lectures on Aesthetics*, I, p. 592–593.

25 To the degree that pathos is generic to character that is artistically construable, the lack of pathos in the agency depicted in this third form of romanticism renders it problematic for art.
26 Hegel, *Lectures on Aesthetics*, I, p. 577.
27 Hegel characterizes these two alternative configurations of the formal independence of character as "formal firmness of character" and "character as inner but undeveloped totality", respectively. See Hegel, *Lectures on Aesthetics*, I, pp. 577–584.
28 Hegel, *Lectures on Aesthetics*, I, pp. 577–580.
29 Hegel, *Lectures on Aesthetics*, I, p. 580.
30 Hegel, *Lectures on Aesthetics*, I, p. 581.
31 Hegel, *Lectures on Aesthetics*, I, pp. 583, 585.
32 Hegel, *Lectures on Aesthetics*, I, p. 575.
33 Hegel, *Lectures on Aesthetics*, I, p. 590.
34 Hegel, *Lectures on Aesthetics*, I, pp. 573ff.
35 Hegel, *Lectures on Aesthetics*, I, pp. 573–574, 595–600, 600–602. He does, however, describe these as aspects of the dissolution of romantic art. (Ibid., p. 595.)
36 Hegel, *Lectures on Aesthetics*, I, pp. 595–596.
37 Hegel, *Lectures on Aesthetics*, I, p. 596.
38 Hegel characterizes early modern Dutch art in these terms. See Hegel, *Lectures on Aesthetics*, I, pp. 597–598.
39 Hegel, *Lectures on Aesthetics*, I, pp. 601–602.
40 This might seem to be the one case where Kant's account of aesthetic judgment might appear to be applicable. However, even here a cognitive element must be paid heed, for unless the audience comprehends that the work's appearance is to be understood as an affirmation of the independent subjectivity from which it originates and to which it is manifest, art succumbs to blind sensation.
41 Since the determination of artforms occurs independently of that of the individual arts, non-objective art cannot here signify simply a form of visual art. It must rather apply to all individual arts, provided that they have the resources for bringing to expression a retreat from the external world and reconstitution of a purely subjective domain of appearance.
42 Although Hegel does not explicitly address abstract art, the latter can easily be characterized in terms of his treatment of the romantic subjective mastery of content, as has been done here. See Hegel, *Lectures on Aesthetics*, I, pp. 599–600.
43 Because such "formal independence of individual particularities" can be seen in all the above cases, Szondi is wrong to identify this third stage in Hegel's account of romanticism as nothing but realism. (See Szondi, *Poetik und Geschichtsphilosophie* I, p. 440.) It does encompass realism, but equally abstract expressionism, serialism, and other varieties of art in which inwardness exhibits its independent mastery over external configurations.
44 Peter Szondi discusses the significance of this connection for the development of art since Hegel's time. See Szondi, *Poetik und Geschichtsphilosophie* I, pp. 438–440.
45 Hegel's account falls short by failing to take into account non-objective art.
46 Peter Szondi objects that Hegel's account of romantic art leaves out a fourth variety, which Szondi identifies as "modern" art, a variety of which Hegel could have no knowledge since it allegedly comprises the type of art that continued to be produced in the century and a half since Hegel's death. (Szondi, *Poetik und Geschichtsphilosophie* I, pp. 461–462.) However, as Szondi himself ends up suggesting (ibid., pp. 464–465), none of the "post-Hegelian" artistic movements to which he refers as exemplars of this fourth type fall beyond the polar extremes of realism and non-objectivism defining the boundaries generic to the third form of romantic art. Further, since these forms are conceptually distinguished, their applicability is not historically relative in the way in which empirically derived models would be.

47 Hegel, *Lectures on Aesthetics*, I, p. 606.
48 Hegel, *Lectures on Aesthetics*, I, p. 607.
49 Arthur C. Danto, *The Transfiguration of the Commonplace* (Cambridge, MA: Harvard University Press, 1981), pp. vii, 56.
50 Hegel, *Lectures on Aesthetics*, I, pp. 607, 609.
51 The problem is not that Hegel's unconditioned claims about historical necessity are undermined by the alleged fact that he, like all purveyors of a philosophy of history, are bound to the standpoint of their own age, so that his conception of historical development turns into a self-understanding of one age through the appropriation of past eras. Such an argument, advanced in relation to Hegel's *Lectures on Aesthetics* by Peter Szondi (see Szondi, *Poetik und Geschichtsphilosophie* I, 433) is undermined by its own unconditioned global claims about the historicity of any philosophy of history.
52 Hegel, *Lectures on Aesthetics*, I, p. 594.
53 Hegel, *Lectures on Aesthetics*, I, p. 594.
54 Hegel, *Lectures on Aesthetics*, I, pp. 528–529.
55 Hegel, *Lectures on Aesthetics*, I, p. 575.
56 Hegel, *Lectures on Aesthetics*, I, pp. 576, 595, 602.
57 Hegel, *Lectures on Aesthetics*, I, p. 575.
58 Hegel, *Lectures on Aesthetics*, I, p. 596.
59 Hegel, *Lectures on Aesthetics*, I, p. 602.

Seventeen

The Challenge of Architecture to Systematic Aesthetics

Architecture as an Aesthetic Enigma

More than any of the traditional fine arts, architecture presents a puzzle to aesthetics. On the one hand, architecture ordinarily does not denote anything at all. Like music and dance devoid of program and abstract visual art, architecture seems to require that beauty shed the last remnants of mimesis and every other form of representation, yet still provide whatever meaning art should possess. On the other hand, architecture departs from other non-representational arts by ordinarily serving a prosaic purpose, essentially housing some activity or artifact, instead of standing sheltered from use as something whose sheer appearance can independently command attention. How then can architecture be accommodated by the philosophy of fine art, without subverting the significance and autonomy of beauty?

This problem takes an acute form for Hegel, even though he repudiates mimesis and any notion of art for art's sake that detaches aesthetic form from content. More than any other past aesthetician, Hegel recognizes that aesthetic phenomena are pervaded by an individuality tying shape and meaning inextricably together. This ubiquitous individuality is exhibited first in how the beauty of an artwork is dependent upon the concrete details of its execution, rather than merely upon its embodiment of some universal form, secondly, in how artistic production subordinates the universal routines of technique to the unique originality of artistic genius, and finally in how art criticism loses the aesthetic dimension as soon as it judges its object against independently given standards.[1] Despite the aesthetic centrality of beauty's individuality, it escapes the conceptual resources of mimetic theory, pioneered by Plato and Aristotle, and the transcendental aesthetics of reception, inaugurated by Hume and Kant. Both aesthetics fall short of their prey by conceiving beauty to be determined by antecedent factors that inform it with a characteristic universal identity whose particular embodiment becomes incidental to its aesthetic worth. Just as mimetic art is defined by giving expression to the antecedently given form it represents, so transcendental aesthetics derives beauty from the structure of aesthetic reception that underlies each and every aesthetic appearance. In each case, form and content fall asunder, for what individuates the object of beauty from other bearers of the same form or other products of an analogous transcendental constitution drops out of aesthetic consideration.[2]

Because Hegel makes the individuality of beauty central to aesthetics, he realizes that art cannot simply *re-present* given contents, but must *transfigure* whatever it presents so that the form of its appearance is uniquely tied to its content.

Consequently, Hegel's departure from mimesis does not entail an exclusion of meaning from beauty; rather it requires locating those meanings susceptible of artistic construal, a construal whose sensuous individuality needs an equally concrete significance to enable form and content to fit. Moreover, because the unity of form and content signifies that beauty be independent of external factors, the meaning of art must inherently deserve display. Instead of depending upon contingent interests and perspectives, artistic meaning must be of universal significance to any possible audience. Hence, the concrete meaning of art must involve what is of fundamental importance to rational agents per se. Although, besides art itself, ethics, religion, and philosophy each have an unconditioned significance, only ethics and religion have concrete dimensions susceptible of sensuous presentation. Accordingly, when beauty unites form and content, the content will express fundamental truths of ethics or religion, providing rational agents with a self-understanding no matter what their personal experiences or cultural community. Moreover, to the degree that ethics and religion have conceptually determinate alternative configurations, beauty will divide into fundamental artistic styles, each comprising the particular mode of artistic construal by which a particular ethical and religious world view obtains a congruent aesthetic form.

These cardinal desiderata of Hegel's aesthetics magnify the problem of architecture. For not only must architecture somehow retain significance without mimesis, but it must express a meaning fit for art, e.g. a fundamental self-understanding of humanity in a unique sensuous configuration. Consequently, the problem is hardly solved by Nelson Goodman's observation that architecture can refer without denotation by literally or metaphorically expressing contents that may have their own mediated references.[3] The identification of these alternative types of reference still leaves unexplained how architecture can signify a worthy content in a distinctively aesthetic way. Moreover, architecture must retain its aesthetic individuality even while serving a function, whose instrumentality fits the very categories of artisan production that otherwise condemn art to the embodiment of antecedent forms, where the individuality of the artifact is incidental to its functionality. Can architecture be non-representational and functional, yet sensuously convey fundamental human truths with aesthetic individuality? Meeting this challenge is difficult enough, but Hegel's account adds one further complication. Because ethical and religious world views have conceptually determinate varieties entailing particular modes of artistic configuration, an account of architecture must show how these a priori art forms are architecturally realizable.

Hegel does not flinch from meeting these challenges and few would dispute that his treatment of architecture is without parallel for its conceptual scope and ambition.[4] Yet has Hegel succeeded in deciphering the riddle of architecture? The answer to this question dovetails with whether Hegel has succeeded in bringing aesthetics to its third and final level of determination: the theory of the individual arts, which incorporates, and thereby presupposes, the concepts of art in general and of the particular art forms.

Hegel himself identifies architecture as the keystone of the individual arts, preceding all others both in historical origin and in conceptual ordering. If this claim is taken seriously, Hegel's conception of architecture will make or break the entire edifice that is built upon it.

The Principal Suspect Features of Hegel's Concept of Architecture

Almost immediately, Hegel's account of architecture exhibits salient suspect features that persist throughout his treatment of the other arts, calling into question the whole enterprise of his theory of the individual arts.

First, Hegel connects the historical and conceptual primacy of architecture with the symbolic art form, suggesting that architecture is preeminently a symbolic art, limited by its arrangement of heavy masses to relate meaning and shape in the allusive manner defining the discrepancy at the heart of the symbolic art form.[5] Admittedly, if architecture is essentially restricted to erecting immobile, solid masses in conformity with laws of gravity, it cannot present configurations that have a subjectivity of their own. Nevertheless, this limitation hardly prevents architecture from creating environments for sacred and profane conduct expressing markedly different world views, each standing in a different relation to the shape that makes it manifest. These possibilities are symptomatic of a basic conceptual confusion lingering in Hegel's subsequent characterization of sculpture as a classical art and of painting, music and literature as romantic arts. In each case, Hegel seems to conflate two distinct levels of aesthetic determination: that of the particular art forms and that of the individual arts. Instead of treating the art forms as particular modes of relating meaning and shape that are determinable in their own right, independently of the individual arts, Hegel here seems to commit the category mistake of attaching particular art forms to different individual arts, blurring the distinction between the two dimensions of aesthetic reality.

Yet, at the very same time, Hegel cuts his nose to spite his face, developing architecture into a succession of symbolic, classical and romantic developments, demonstrating that architecture can exhibit every art form, just as sculpture, painting, music and literature will prove to do in his own subsequent treatment.[6] Indeed, he even emphasizes that the three art forms play a greater determining role in architecture than in the other arts, whose stylistic differentiations are comparatively limited due to a more encompassing penetration by one art form.[7] At most, Hegel's linkage of architecture to the symbolic art form might support the prescriptive judgment that symbolic architecture will enjoy an artistic success outstripping classical and romantic constructions. Hegel's own analyses of the Greek temple and the Gothic cathedral, however, give the lie to any such verdict, testifying to how artworks of any medium and of any style can rise to aesthetic greatness or sink into banality. For as Hegel's own account of symbolic and romantic art already established, even when meaning and shape are bound to be incongruent, beauty can be attained by making that incongruence as much a part of artistic form as an aspect of artistic content. Similarly, an art form may have greater or lesser privilege as a vehicle for expressing human self-understanding, but so long as it qualifies as art, it gives expression to an essential world view in a sensuous image linked to the meaning it conveys. Accordingly, the very categories of Hegel's aesthetics provide ample reason to minimize the linkage of architecture with symbolic art and to concentrate instead upon investigating the success with which his own development of symbolic, classical and romantic architecture brings to a first concrete realization how the universal and particular categories of aesthetics inform an individual art.

It is here, in the actual working out of Hegel's concept of architecture, that a second set of perplexities arise, revolving around how the instrumentality of buildings enters into the development of architectural style. Hegel claims that architectural style begins, both historically and conceptually, with a symbolic construal where architecture does *not* serve any external use, but constitutes a building to be independently beheld like a mammoth, inchoate, inorganic sculpture.[8] Then, by way of transitional constructions like the Pyramids, where a purpose becomes the hidden core of a monument,[9] a classical architecture arises in which form and function are united, producing buildings that serve as means to an end lying outside them. Finally, architecture assumes a third, romantic form, uniting the defining aspects of its forerunners, exhibiting an independence of form from function while still serving a purpose.[10] Admittedly, this three-fold differentiation of architecture possesses a logical closure by exhausting the basic ways in which building can relate to external purpose: an edifice can either be independent, purposive, or both. What is less evident is how these three relationships of architectural form and function have anything to do with how symbolic, classical and romantic styles achieve architectural realization. Why should an independent architecture be mandated by the symbolic construal of a human self-understanding too abstract to have any adequate sensuous embodiment? How can a union of architectural form and function serve the classical ideal of a human spirit whose substance is as it appears? And why should a unification of architectural function and independence be required for the romantic expression of a subjectivity whose essential inward autonomy can never be fully captured in any external manifestation? Has Hegel here inadvertently testified to how the categories of art in general and the particular art forms cannot lay hold of an individual artistic medium, handing over analysis of the individual arts to art historians and art critics? Or does he pave the way for conceiving a priori forms of architectural style?

The Puzzle of Independent, Symbolic Architecture

Hegel's first candidate in the division of architecture is most puzzling, since the very independence by which it is distinguished seems to transgress the bounds of architecture. As Hegel himself observes, an architecture that builds structures devoid of use verges on monumental, inorganic sculpture.[11] For if a construction shapes heavy masses without regard for some external purpose, how can it be distinguished from a work of sculpture that equally shapes matter purely in regard to its three dimensional form? Hegel might appear to have an easy escape from this problem, given his limitation of genuine sculpture to the plastic shaping of the human form. Granted this restriction, independent architecture and abstract statuary would both fall outside sculpture proper into the domain of architectural design, provided independent architecture divested itself of any representations of the human form.[12] Hegel, however, himself refers to works of independent, symbolic architecture, such as the Egyptian Sphinxes and Memnons, that represent human figures. In these cases, what allows otherwise human shapes to fall within architecture is the monumentality of scale and placement in arrangements whose non-representational plan is the real focal point of architectural interest.[13] For these reasons, Hegel is content to distinguish independent architecture from sculpture, despite the absence of an external function.

The question remains, nonetheless, how the independence of such architecture fits its symbolic style. A plausible answer that seems to fit Hegel's treatment of classical and romantic architecture, runs as follows: the restriction of architecture to the arrangement of heavy masses prevents buildings from adequately conveying the subjectivity and autonomy of rational agency. Hence, when that subjectivity and autonomy is understood to be of fundamental significance, architecture can serve the ends of art only by being a means for housing either artworks in other media more capable of presenting the truth of independent agency (such as sculpture and painting) or for housing activity in which rational agency is properly realized. When, by contrast, the self-understanding of humanity rejects the affirmative significance of human finitude and its independent subjectivity, and instead places ultimate value in divine powers into which individuality is absorbed and extinguished as an illusory epiphenomenon, the truth of existence can be directly conveyed with the abstract resources of architecture. In the former case, which fits the world views of classical and romantic styles, architecture retains its functionality. In the latter case, which fits the self-understanding expressed by symbolic art, architecture can be independent, simply making manifest the truth to which its static, inorganic forms can so suitably allude.

Such an explanation may account for the linkage between independence and symbolic architecture and between functionality and classical and romantic architecture, but these connections still leave undetermined how the respective independence and dependence of symbolic and non-symbolic building has ramifications for the specific appearance of these architectural styles. After all, whether a building is constructed for its own sake or for an external use need not automatically affect its design in any fashion whatsoever. The same plan could conceivably fit either option.

Hegel himself seems to admit the indeterminacy of the connection between the independence of symbolic architecture and architectural design when he observes that the variety of both form and content in symbolic architecture can be neither exhausted nor systematically presented.[14] He attributes this to the absence of any set content or mode of design, an absence reflecting how the meanings in symbolic art are abstract universal representations, whose unformed generality fits no sensuous image, but gains expression only by drawing in an arbitrary, haphazard fashion from elementary physical processes, animal and plant life, and aspects of spiritual reality, never achieving unity in an individual subjectivity.[15] Nevertheless, Hegel proceeds to distinguish three forms of independent architecture in an attempt to bring a rational division to the most important options of symbolic construction.[16] These three forms consist first of buildings that unify a nation,[17] both in the task of construction and, upon completion, by providing a focal point of gathering; secondly, of buildings in which symbolic meanings differentiate themselves in inorganic and organic shapes; and finally in transitional buildings in which sculptural configuration is excluded from the edifice that now provides enclosure for meanings not directly expressed in its architecture.

The first and third forms offer little towards exemplifying how the meaning tied to the independence of symbolic architecture has any ramifications for architectural design. Buildings identified by their role in unifying a nation seem all too burdened by an external purpose and all too undetermined in respect to shape. If it is the construction of an edifice that unifies a people, as Hegel alleges was the case with the tower of Babylon,[18] the appearance of the completed work is merely a superfluous outcome. If, alternately, the finished work unifies the nation by being a place in which people gather in confirmation of national unity, the achievement of this function seems unrelated to any specific architectural configuration, for as Hegel himself observes, the form and shape of such a building can only express its unifying function in an external way.[19] By contrast, the third, transitional form of "independent" architecture may well entail the sort of design features Hegel finds exemplified in the Pyramids, whose geometric severity excludes sculptural detail and whose hidden burial chambers serve a significance not manifest on the exterior. Nonetheless, the reference to an end external to the architectural design risks transgressing the independence of independent architecture, even if the concealing and sealing of the interior passages prevents the living from making any use of what lies within.

The second form of independent architecture more squarely offers the sought connection between symbolic meaning and specific building designs. Here, the abstract natural forces in which symbolic meaning invests absolute worth gain architectural expression in such independent monuments as obelisks and lingams, whose concrete forms allude to those powers. Alternately, symbolic architecture erects monuments in the mold of organic forms whose transgression of natural shape and dimensions and whose combination of human and animal features express the sublimity of a divine in which independent subjectivity is overcome. As Hegel admits, symbolic architecture here veers closest to figurative sculpture. Nonetheless, it remains an art of building by, on the one hand, enlarging its figures to colossal proportions that render them lifeless masses, and, on the other hand, placing them in arrangements that make them elements of an inorganic architectural scheme.[20] Although Hegel cites historical examples for each option, these figure as instances of a type of construction that can be conceptually determined: phallus columns and obelisks are representative of non-figurative, inorganically shaped constructions that have no use other than symbolizing abstract natural powers just as Memnons and Sphinxes are free standing organic monuments, whose mixture of animal and human features, reduction of the individuality of the figure, and monumentality and schematic placement serve only to allude to divinities for whom individual subjectivity has no independent stature.[21]

If these stylistic options provide a connection between architectural independence and symbolic meaning, the question still arises as to whether a symbolic architecture can construct temples, palaces, and other useful buildings without transgressing its own self-understanding and passing beyond into a classical or romantic mold, where, if Hegel is to be followed, the instrumentality of architecture reflects recognition of the centrality of human individuality. Such possibilities need not be denied, for buildings serving human needs can equally express symbolic art's abstract notions of divinity and self provided they possess the same modalities of design whose independent embodiment has just been described. Indeed, the very examples of symbolic architecture to which Hegel refers combine their obelisks, lingams, Memnons and Sphinxes in architectural complexes in which use is not ignored, but in which form and function are not yet united.

The Unity of Form and Function in Classical Architecture

Symbolic architecture may thus retain an independence in keeping with the world view it architecturally expresses, but can architecture embody the classical ideal by uniting form and function? Hegel's prescription for an architecture that employs rectilinear geometric form to enclose a form of life sure of its own independent humanity almost seems to converge with slogans of Bauhaus functionality, Le Corbusier's manifesto for a new architecture in the instrumental spirit of modern technology,[22] and Frank Lloyd Wright's call for a natural house fitting human dimensions by excluding superfluous ornament and shape.[23] Can these brands of architectural modernism avoid conflation with classicism and somehow retain a place within romantic architecture, where functionality and independence go hand in hand? Or is the identity of classical architecture too ill-defined to save architectural development from wholesale confusion? To resolve these questions, the unity of form and function must be further specified in regard to both the function at issue and the form that fits.

Whether Hegel succeeds in identifying a conceptually distinctive unity of meaning and configuration for classical architecture must be judged from his analysis of the Greek temple, which forms the crux of his treatment. As a realization of the classical style, the Greek temple cannot serve simply as an enclosure for worship. That function is too broad, since any religious building could share that purpose, no matter what architectural style it possessed. Instead, the temple's function must be more narrowly identified in respect to both the anthropomorphic polytheism characterizing the religious self-understanding of classicism and the affirmation of a naturally defined, but independent agency, for whom everything of worth lies in the outer activity of individuals, whose sensuous appearance is sufficient to convey their essential significance. Classical religious representation, unlike its symbolic counterpart, can find a perfect expression of what is divine in the figure of finite rational agency (e.g., in our terrestrial terms, the human body), just as the public life of individuals can be a sufficient arena for revealing everything essential in human existence. If classical architecture expresses a world view with these features, and classical form follows function, the design of the classical temple must distinctively serve them.

Since the classical divine can be adequately represented in the figure of finite rational agents, figurative arts can be relied upon to convey the presence of divinity within the temple precincts. To the degree that sculpture presents only the surface of its figures, without supplying the idealization that two-dimensional figurative art adds, statues of the anthropomorphic gods would readily fit the demands of representing a form of agency that exhausts its fundamental significance in its outer deeds. Moreover, since such divinities have freed themselves from subordination to natural powers, they find their own suitable element only in an inorganic environment produced by rational artifice.[24] Accordingly, the classical temple can serve the function of enclosing statues of such divinities, together with the activity by which classical worshipers are distinguished. Since the classical world view invests ultimate significance in the worldly, sensuously manifest, public display of autonomous individuals, the classical temple will not shut off the religious community from the secular public world. Instead of enclosing the place of worship in an ethereal chamber, secluded from profane life, the classical temple will be open to the outside, permitting worship to coexist with the conduct of public life.

When architectural form follows these specific functional considerations, salient features of the Greek temple readily result. To begin with, if building is to restrict itself to housing a type of divinity and rational agency that is adequately represented by idealized human statues and by the public conduct of the members of a naturally defined ethical community, the stylistic signatures of symbolic architecture must be shed.[25] Independent inorganic shapes, alluding to abstract divinities in which individuality is absorbed, have no more a role to play than plant, animal and spiritual figures and motifs that symbolize sublime forces transcending any given sensuous, let alone human, embodiment. Instead, the edifice must reduce itself to a purely instrumental enclosure for what transparently makes manifest its own significance.[26] In this way, form and function are united, with architectural design governed by the demands of supporting an enclosure that will at once contain statuary of human figures and enable worshipers to view the representation of their anthropomorphic deities without leaving the public world behind.[27] As Hegel painstakingly argues, the classical column provides precisely the architectural element for achieving this dual aim. With its cylindrical base, the column offers support while remaining a free standing element that provides a space for worship without cutting off the outside scenery, as a continuous wall would do.[28] By progressively narrowing as it ascends, the column incorporates in its profile the progressively diminishing burden of support, just as the spacing and thickness of columns precisely reflects their function by being no more frequent and wide than necessary to support the load above.[29] Similarly, the addition of base and capital brings functional closure to the column, giving it a height determined by the needs of support, in contrast to the indefinite reach of a post, whose subterranean depths cannot be fathomed from what juts above the surface.[30] Finally, the roof supported by the rows of columns fits its purely sheltering role by coming to a point, upon which nothing more can find support, unlike a horizontal roof, whose capacity to bear further stories renders it an arbitrary summit to a building that might as well continue to ascend.[31] Here the needs of enclosing anthropomorphic statuary and of coming and going between worship and public life precludes any upward striving; instead, the facade extends primarily in width, allowing the eye to take in the design without having to gaze upwards.[32] In these ways, the rows of columns get connected by horizontal beams, bearing friezes and cornices underlying a peaked roof, creating a unified structure whose design neatly fits the functional demands of the classical ideal. Although the resulting temple may approach the geometric severity of modernist fashion, the open rows of tapering columns and the peaked roof have little in common with the reflecting glass facades and barren plazas of our unadorned Bauhaus boxes. The fit of form and function in classical architecture is predicated upon a very particular purpose that only the self-understanding of a substantial subjectivity can provide: the housing of a rational agency whose outward appearance is at one with its essence.

The Unity of Independence and Functionality in Romantic Architecture

If the anatomy of the Greek temple testifies to how unity of form and function gives architectural expression to the classical ideal, can any example of romantic architecture confirm Hegel's claim that a unity of independence and function architecturally expresses the self-understanding of romanticism? Hegel's own treatment of romantic architecture offers only one candidate for vindication: the Gothic cathedral. Although Hegel's general concept of the romantic style divides into three principal developments (one configuring the religious self-understanding generic to Christianity, another portraying the adventures of chivalry, and the last expressing a formally independent character), his analysis of romantic architecture focuses almost solely upon the architectural equivalent of the first option.[33]

In contrast to classical temple construction, romantic architecture must somehow give shape to a religious self-understanding that upholds the infinite worth of a free agency whose independent subjective life cannot find adequate expression in any given appearance, including the exterior of the human body and the purely public face of conduct. The inwardness of each self must here finally find due recognition precisely by transcending each and every particular sensuous embodiment. Consequently, romantic architecture cannot simply be an enclosure for statuary whose bodily surface independently expresses everything of spiritual importance, nor can a romantic place of worship remain open to the scenes of secular life, as if the public appearance of individuals exhausted their integrity.

Hegel maintains that the Gothic cathedral provides the positive architectural answer to these injunctions in several salient ways.

First, the Gothic cathedral provides a fully enclosed space, explicitly severed from the activities of mundane secular life.[34] Whereas the classical temple can join exterior and interior through rows of supporting columns and open cells, the Gothic cathedral blocks out the outer world by all-encompassing walls,[35] punctuated by curtains of stained glass admitting a light distinctly different from prosaic illumination.[36] Moreover, the exterior of the building expressly reflects the totally enclosing design of the interior, emphasizing the architectural priority of what lies sequestered within over external nature and the finite involvements of secular life outside.[37]

Secondly, instead of employing free-standing columns that support a roof just high enough to serve the needs of displaying anthropomorphic sculpture, the Gothic cathedral encloses its interior space so as to mask the supporting role of its walls, transforming them instead into vehicles of an immense elevation above and beyond any merely rectilinear functional requirement.[38] This transfiguration is achieved in two ways. On the one hand, in place of columns limited by base, capital and cross beam, the Gothic cathedral employs immense pillars, divided into ascending tendrils that meet at the apex of a soaring ceiling.[39] The resulting pointed vaults and arches overcome the perpendicular architectural transition from supporting walls to supported roof, seamlessly joining pillar and peak,[40] while forcing the eyes of all visitors to turn upwards in order to take in the whole view.[41] On the other hand, the soaring walls are pierced and punctuated by vast tapestries of stained glass and ornament that dissolve their load-bearing solidity, suggesting a lightness in keeping with their elevating role.[42]

Through these signature design innovations, the Gothic cathedral provides an environment that fits a spirituality that looks beyond the external sensuous realizations of humanity to an interiority available to all, who literally have space to fit within an inner sanctum sufficiently immense to accommodate all walks of life of the local populace.[43]

Yet if Hegel here succeeds in matching Gothic style with the religious self-understanding of romanticism, has he confirmed the unification of independence and functionality by which romantic architecture was to have exhausted the logical options of architectural design? Functionality is certainly at hand insofar as the Gothic cathedral serves to house the worship of a religious community, instead of standing alone as an independent monument, to be appreciated merely for its allusive appearance. Yet independence is equally present insofar as the design does not simply merge form and function, restricting itself to the geometric simplicities mandated by support and enclosure. The soaring height, the vaults and arches of merging pillars, and the curtains of colored glass and ornament all take on a meaning of their own, congruent with the religious practice they surround, but transcending the practical requirements of worship.[44]

Can the same be said of modernist and "post-modern" architecture, which Hegel never addressed, but which his stylistic scheme would have to place within the orbit of romantic architecture?

The modernist call to liberate architecture of sclerotic ornament and traditional design might appear to be a reversion to the fit of form and function characterizing classical architecture. Whereas, however, classicism could treat architecture as a stage for a humanity fully manifest in its bodily, public appearance, modernism seeks to provide an environment for a humanity to whom no given order can be the final measure of its autonomy.[45] The modernist turn to spare geometric forms is not intended as a backdrop for human meanings requiring a figurative expression, but rather as an independent manifestation of the transcendence of given tradition, where Bauhaus minimalism architecturally configures the same self-understanding as abstract art and atonal music, a self-understanding that confers ultimate significance to a rational agency that avoids reproducing what is already given in nature or tradition by attending to what is merely the product of its own creativity.

"Post-modern" architecture offers a variation on this modernist agenda by ironically quoting past styles as if to show how all stylistic traditions are relegated by their historicity to be mere playthings of the artistic self, for whom no external embodiment can be definitively adequate.[46] Although post-modernism rejects the purified abstractions of modernism as but another stipulated "essentialism" of one more vain tradition, the post-modern quotational distancing from every determinate style only reaffirms, albeit in a perilously formal way, the modern overcoming of foundations and embrace of self-determination as the unique foundation-free alternative to tradition.[47] Once more, building composition transcends the practical use to which construction is put, since here every design can just as well be cast aside and replaced by some other quotation. Whether repudiating all traditional conventions by advancing abstract schemes or acknowledging traditions' relativity by arbitrarily recycling their elements, the architecture of our day readily embodies the shapes of romanticism that Hegel identifies with the independent formal character possessed by the modern self. By recognizing the universal validity of free agency, modern humanity can no longer invest any particular sensuous configuration with ultimate meaning. For architecture, this signifies that building design must create novel works that break free of each particular mold, housing the activities of a free community in structures whose architectural form is liberated from both total subordination to function and slavish reworking of established modes. Whether the Wrights and Le Corbusiers of tomorrow can ever overcome the restlessness of their vocation will depend upon the durability of the institutions and culture of freedom. In providing for this undecided prospect, Hegel's aesthetics proves to have met the challenge of architecture.

*****

Although the connection of autonomy and normativity is as central to aesthetic worth as it is to truth and right, modernity's overcoming of foundations leaves art with a more ambiguous prospect than that of theory or practice. To leave behind the complementary limitations of the Enlightenment and post-modernism, to secure the legitimacy of modernity, and to recover the normativity of truth, right and beauty with due recognition of the respective situations of philosophy, conduct and art, is the task now to be concluded.

Notes

1 Michael B. Foster succinctly makes these points in *The Political Philosophies of Plato and Hegel* (New York: Oxford University Press, 1967), pp.185–87, 201.

2 For a detailed exposure of these limitations of mimetic, e.g. metaphysical, and transcendental aesthetics, see Richard Dien Winfield, *Systematic Aesthetics* (Gainesville, FL: University Press of Florida, 1995), Part 1, pp. 15–57.

3 Nelson Goodman, "How Buildings Mean", in *Aesthetics: A Critical Anthology*, edited by George Dickie, Richard Sclafani and Ronald Roblin (New York: St. Martin's Press, 1989), pp. 544–555.

4 The contrast between Hegel's account of architecture and Vitruvius' celebrated *The Ten Books of Architecture* is indicative of the gap between Hegel and the competition. Vitruvius totally ignores the connection between meaning and configuration in architecture, offering only vague notions of harmony and proportion that provide little insight into the distinctive significance and beauty of the Greek temple on which he focuses his attention. See Vitruvius, *The Ten Books of Architecture*, trans. by Morris Hicky Morgan (New York: Dover Publications, 1960).

5 Hegel, G.W.F., *Vorlesungen über die Ästhetik*, ed. E. Moldenhauer and K. Michel, *Werke in zwanzig Bänden*, Vol. 14 (Frankfurt am Main: Suhrkamp Verlag, 1970), p. 269.

6 This is true even though Hegel maintains that literature is divided into epic, lyric and dramatic forms, instead of into embodiments of the three art forms (*Ästhetik, Werke 14*, p. 271). As Georg Lukács, Peter Szondi and others have shown, such literary genres either fit a particular art form or undergo developments in which all art forms are represented. Thus, for example, Lukács argues in his *The Theory of the Novel*, trans. by Anna Bostock (Cambridge: MIT Press, 1971), that the epic is a preeminently classical form of literature whereas the novel is a preeminently romantic literary genre, whereas Peter Szondi argues, in his *The Theory of the Modern Drama*, trans. by Michael Hays (Minneapolis: University of Minnesota Press, 1987), that drama undergoes a development in which romantic experiments supplant classical models.

7 *Ästhetik, Werke 14*, p. 271.
8 *Ästhetik, Werke 14*, p. 270.
9 The Pyramids are *only* transitional constructions because of key limitations in their unity of form and function: although pyramids serve as enclosures of the dead, using inorganic shapes that are expressly artificial forms, built by and for humanity, the failure to use right angles as a ruling principle gives the structure a tapering closure that is not useful to the living. See *Ästhetik, Werke 14*, p. 296.
10 *Ästhetik, Werke 14*, p.270.
11 *Ästhetik, Werke 14*, p. 270.
12 In this vein, Hegel observes that although buildings without any distinction in the aim of enclosure and the building as fulfillment of this aim would be like works of sculpture, such independent architecture remains differentiated from sculpture by not producing images that are themselves spiritual and subjective. See *Ästhetik, Werke 14*, p. 269.
13 *Ästhetik, Werke 14*, pp. 279, 282–3.
14 *Ästhetik, Werke 14*, pp. 274–5.
15 *Ästhetik, Werke 14*, p. 274.
16 *Ästhetik, Werke 14*, pp. 275–6.
17 Hegel here anticipates something of Wittfogel's notion of oriental hydraulic civilizations, where national unity centers around vast construction projects (in Wittfogel's case, of irrigation systems and the like). See Wittfogel, Karl August, *Oriental Despotism: A Comparative Study of Total Power* (New Haven: Yale University Press, 1957).
18 *Ästhetik, Werke 14*, p. 276.
19 *Ästhetik, Werke 14*, p. 277.
20 *Ästhetik, Werke 14*, pp. 275, 279.
21 *Ästhetik, Werke 14*, pp. 280–3.
22 For Le Corbusier's classic statement of this injunction, see his *Towards A New Architecture*, trans. by Frederick Etchells (New York: Dover Publications, 1986).
23 See Frank Lloyd Wright's *The Natural House* in Frank Lloyd Wright, *Collected Writings, Volume 5: 1949–1959*, ed. by Bruce Brooks Pfeiffer (New York: Rizzoli International, 1995), pp. 77–127.
24 *Ästhetik, Werke 14*, p. 297.
25 *Ästhetik, Werke 14*, pp. 302–3.
26 *Ästhetik, Werke 14*, p. 304. As Le Corbusier analogously expresses it, describing the Parthenon, "no symbolical description, no naturalistic representation; there is nothing but pure forms in precise relationships". See Le Corbusier, *Towards A New Architecture*, p. 220.
27 Le Corbusier expresses an analogous appreciation of this classical departure from symbolic architecture, by observing, in regard to the Parthenon, how "We are in the inexorable realm of the mechanical. There are no symbols attached to these forms." See Le Corbusier, *Towards A New Architecture*, p. 211.
28 *Ästhetik, Werke 14*, pp. 299, 310.
29 *Ästhetik, Werke 14*, pp. 299, 306, 310, 312.
30 *Ästhetik, Werke 14*, pp. 310–11.
31 *Ästhetik, Werke 14*, pp. 314–5.
32 *Ästhetik, Werke 14*, pp. 318–9.
33 Although Hegel briefly mentions medieval secular architecture, he brushes it aside, noting that the walls, towers and bridges of its castle-like fortifications were so governed by utilitarian demands that artistic considerations only entered in as incidental decoration. See *Ästhetik, Werke 14*, p. 348.
34 *Ästhetik, Werke 14*, p. 332.

35 Hegel thus identifies the wall as the prime element in romantic construction. See *Ästhetik, Werke 14*, p. 310.
36 *Ästhetik, Werke 14*, p. 333.
37 *Ästhetik, Werke 14*, pp. 334, 340, 342.
38 *Ästhetik, Werke 14*, p. 335.
39 *Ästhetik, Werke 14*, pp. 336–7.
40 *Ästhetik, Werke 14*, pp. 333–4.
41 *Ästhetik, Werke 14*, p. 337.
42 *Ästhetik, Werke 14*, pp. 338, 345.
43 *Ästhetik, Werke 14*, pp. 339–40.
44 Le Corbusier acknowledges this independence when he describes the Gothic cathedral as "A fight against the force of gravity, which is a sensation of a sentimental nature". See Le Corbusier, *Towards A New Architecture*, p. 30.
45 Le Corbusier's "house-machine" is thus hardly congruent with the instrumentality of the Greek temple. See Le Corbusier, *Towards A New Architecture*, p. 7.
46 David Kolb persuasively argues how the "ironic historicism" of post-modern architecture, despite its own claims to the contrary, is merely an extension of the same modern self-understanding that motivates modernism's search for pure form. See David Kolb, *Postmodern Sophistications: Philosophy, Architecture, and Tradition* (Chicago: The University of Chicago Press, 1990), p. 87ff.
47 Post-modernism flirts with danger when its ironic play with every tradition turns into a repudiation of the universal validity of self-determination. It then reverts to an embrace of particularism, where the alleged groundlessness of every determinate configuration gives way to a will to power, where no value can be affirmed that is not a perspectival imposition. In this fashion, the post-modern ironic distancing from every tradition becomes an invitation to fascism's promotion of the given particularity of a community over and against every aspiration of universal right. It should thus come as no surprise that the "master thinkers" of post-modernism could welcome Nazism.

Conclusion

Eighteen

Modernity and the Recovery of Truth, Right and Beauty

Truth, right and beauty, once hallmarks of humanity's triumph over meaningless contingency, have become targets of a spreading disdain that threatens the fabric of modern times. Instead of representing the universal, unconditioned measures of knowledge, conduct and art, truth, right and beauty have been reduced to meaning, value and taste, factors relative to particular historical conventions as contingent as any ideals they ground. Cognition, action and artistry are denied the autonomy to provide universally valid achievements, untainted by particular conditions. Condemned to dependency upon historically relative practices, the search for truth, the quest for justice and the striving after beauty either capitulate to the reigning tradition of some local community or revel in their own arbitrariness, imposing openly particular values in exercise of a will to power. The future is made hostage to the past as science, conduct and art bow to common usage or to the bare stipulation of a historically relative self. Yet because cognition is left conditioned by contingent convention, the past forfeits its own objectivity, ceasing to be a topic for impartial understanding and instead becoming an instrument for strategic use, recast according to pragmatic plans for the future.

Modernity and Foundations

Can modernity counter these challenges and uphold its original promise to liberate the future from the past and to liberate the past from the future? In departure from all other forms of civilization, modernity advances the project of erecting a way of life in which nothing can count as legitimate unless it is sanctioned by reason. Tradition, roots, kinship, faith and every other given relationship are called into question. All norms – practical, theoretical and artistic – must first be justified by rational inquiry to warrant compliance. Accordingly, reason must free itself of obedience to external authority, be it sacred or profane, for unless reason can be autonomous, it can hardly be its own judge and achieve the self-legitimation on which reason's own sovereign role depends. Not surprisingly, by placing the autonomy of reason at the pinnacle of authority, modernity ends up extending legitimacy to no other conventions than institutions of freedom. Whether in the household, in society, or in politics, the only practices that modernity recognizes to conform with rational autonomy are those that provide some mode of self-determination.

In face of reigning tradition, the birth of the modern therefore calls for liberating the future from the past. For if right is to prevail and right is the reality of freedom sanctioned by reason, the habits of the past must have no irrevocable hold upon what is to be done. Yet if reason has authority in virtue of its own autonomy and the autonomy of reason is not an empty flight from the given but the fountain of truth, then the past must be liberated from the future, as a reality fit for impartial comprehension independently of any pragmatic aims. Modernity's task is not to forsake comprehending reality in favor of transforming it. The challenge of the modern spirit is to secure objective knowledge and to build a rational second nature in which rights prevail over oppression. Irrational tradition should be overcome, but if reason is to retain its own claims, the past must not be robbed of its independent objectivity.

These agendas endow modernity with a special authority that no passage of time can diminish nor any other form of civilization can rival. Through the revolutionary act of identifying validity with rational autonomy, modernity sets its norms free of the dependence upon privileged foundations that plagues every alternative form of life. If thought and action are not autonomous, they owe their design to some independent factor, which serves as their ground. Any attempt to sanction theory or practice then falls into the dilemma of foundational justification, whose diagnosis is as elusively simple as its cure. So long as legitimacy is sought by appeal to a privileged foundation conferring validity, what possesses validity, the foundationally sanctioned thought or conduct, is something different from the foundation that endows it with validity. This difference guarantees that the source of normativity can never satisfy its own standard for justification. With justification equated with being determined by the privileged foundation, that foundation can only enjoy the same legitimacy it confers if it is grounded upon itself. In that case, however, it would lose its constitutive character as the prior foundation of something grounded, a character depending upon the distinction between ground and grounded. To possess normativity, the source of normativity would instead become something self-grounded. The foundation and its derivative would thereby become identical, supplanting the ground/grounded dichotomy with a unity where what boasts legitimacy is determined by itself. So long, however, as what counts as legitimate is not self-determined, legitimation retains the self-debilitating distinction between founding source and founded recipient that is the fatal bane of every version of foundationalism. This distinction not only robs foundational justification of the ability to achieve coherence, but drives it to eliminate this its own defining distinction once it attempts to become self-referentially consistent.

The resulting solution to the legitimation problem of foundationalism testifies to how the normativity of self-determination is not rooted in any foundation. On the contrary, the normativity of self-determination is the very outcome of foundational justification's own attempt to be consistent and overcome its own incoherence. Instead of depending upon the support of some external criterion, the legitimacy of freedom emerges from the immanent critique of foundationalism, a critique coming to fruition when the calling into question of the validity of any privileged foundation ineluctably entails canceling the distinction between legitimating foundation and the bearer of legitimacy. The freedom from foundations of self-determination thus has a double necessity: not only would self-determined thought, action and art lose their autonomy if their measure lay outside them, but the very appeal to an external standard ends up ratifying the exclusive legitimacy of freedom the moment any such standard is evaluated on its own terms.

These relations define the fundamental divide separating modernity from its opponents as well as the limitations of their opposition. If modernity be the form of civilization liberating itself from foundations by embracing the normativity of rational autonomy, it stands distinguished from pre-modern and post-modern community, the two other basic forms that civilization can have with regard to foundations.

Pre-modernity and Legitimation

Pre-modern civilization takes its bearings by tradition, understanding legitimation to reside in a privileged foundation. Instead of placing theory and practice before the tribunal of autonomous reason, pre-modernity accepts the legitimacy of given custom, privileging the particular heritage in which it locates its roots. Although pre-modern civilization may coexist with or follow the demise of a modern community, pre-modernity is logically antecedent to modernity insofar as modernity emerges by calling tradition into question and struggling to erect institutions realizing the sovereignty of reason and freedom. Because the institutions of freedom are not rooted in any antecedent custom, nor conditioned by anything other than the general prerequisites of rational agency, these modern conventions are potentially global and, perhaps, intergalactic in application. By contrast, because pre-modernity finds legitimacy in the given practices of a particular tradition limited in space and time, it has a local, parochial reach that cannot consistently attain the global hegemony that a rootless modernity can achieve. Further, because the traditions of pre-modernity are not determined by reason and the concept of self-determination, but simply given by the contingencies of convention, pre-modern civilization can possess an inexhaustible variety, whose particulars can be known not by philosophical thought but only by historical cognition.

No matter what shape pre-modernity takes or where and when it appears, it always faces legitimation problems once the sanctity of its privileged practices are questioned. Because they are founded upon a particular tradition, whose authority resides simply in its givenness, each competing heritage has just as arbitrary a standing as any other. Since each tradition is itself a contingent convention whose claim to privilege cannot be rationally justified, nothing but the force of prevailing acceptance can uphold its authority. Once that acceptance begins to crumble under the weight of some alternative, one foundation gets supplanted by another, subject to the same contest for power.

Consequently, when pre-modern movements challenge modernity for violating their cherished traditions, the accusation has force only for those who already accept the authority of pre-modern roots. Even if the pre-modern reaction accuses a modern community of failing to achieve its promised emancipation, that failure cannot sanction pre-modern tradition if it too fails to conform to modern ideals of freedom.

The Post-Modern Challenge to Modernity and the Limits of the Enlightenment

In order for the claims of modernity to be called into question, more must be done than appeal to a privileged tradition or expose the inconsistencies of a particular modern community. Declaiming a discrepancy between modern institutions and some given way of life will not suffice since the privilege of any tradition can always be denied. Unmasking hypocrisies in some modern civilization, on the other hand, still leaves open the possibility that a more consistent modern regime could overcome those duplicities. Accordingly, if modernity is to be contested, the very possibility of fulfilling the modern project must be cast into doubt.

Post-modernity defines itself on the basis of just such suspicion. Although post-modernity recognizes the unjustifiable arbitrariness of every privileged tradition, post-modernity denies that normativity can escape the hold of foundations. Whereas modernity aspires to recognize no other authority than the autonomy of reason, conduct and artistic creativity, post-modernism counters that the alleged independence of rational autonomy is a sham. Like every other candidate for normativity, freedom, post-modernism declares, equally rests upon foundations, whose immediate givenness renders them subject to the same charge of arbitrariness as any alternate privileged given.

Post-modernism finds easy confirmation of this dependence in the Enlightenment's attempt to provide foundations for autonomous reason, right and aesthetic worth. Starting with Descartes, continuing with his empiricist counterparts, and culminating in Kant and successive transcendental philosophers, Enlightenment theorists have pursued the strategy of overcoming dogmatic appeals to the putative givens of nature by turning instead to the structure of cognition as the foundation of objective knowledge. Transforming first philosophy from ontology into epistemology, these theorists supposed that the pitfalls of direct appeal to given reality could be avoided by locating conditions of knowing that determine the character of the objects of knowledge. Rather than contemplating a reality that is simply found, this modern strategy fixes upon the structure of knowing as the foundation from which objectivity is constructed. Yet, as post-modernism can duly observe, this turn involves its own appeal to the given – in this case, the givenness of the structures of knowing by which objectivity is allegedly constructed. These epistemological givens are just as immediately contemplated as the ontological givens that pre-critical metaphysics dogmatically intuited. The basic dilemma of foundationalism applies once more: the conditions of objective knowledge cannot be known to constitute objectivity unless they themselves can be known with the same authority as the knowledge they found. But how can the conditions of knowing be objectively known if they remain antecedent epistemological foundations that must be stipulated *before* the limits of objective knowledge can be determined? The successive turns to the ego, the psychology of human understanding, the transcendental apperception, ordinary and ideal language, the hermeneutic situation, and every other candidate for privileged determiner of objectivity all run up against the same stumbling block – the knowing of the conditions of reference that is to precede objective knowledge stands in need of its own critique. Otherwise every claim about the character and privileged role of the alleged epistemological foundation remains no more than an arbitrary assertion. But how can valid claims be made about the foundations of knowing without taking for granted the adequacy of the cognition that addresses that question? The impasse seems insurmountable. Any attempt to identify foundations for reason is bound to fail.

Completely analogous difficulties plague the Enlightenment strategies to rescue right and beauty from the metaphysical assumptions of tradition. Enlightenment ethics takes its cue from the inability of pre-modern ethics to identify and legitimate a highest good, without which all ends appear relative, condemning conduct to an aimless futility. Every candidate for highest good faces the same legitimation problem as any other privileged foundation – because the highest good is that which confers value upon all other goods, it cannot derive its own value from any other factor. But then nothing but its own given content can be the source of its authority. Yet every other competitor for highest good has a given content of its own that can just as well be immediately affirmed. If no highest good can be justifiably privileged any more than can any first principle of reality, what options does ethics have left? The absence of a highest good certainly sets the will free from any prescribed end, but does this liberation redeem conduct from the aimless quandary of nihilism? The Enlightenment answer is that the collapse of the quest for a highest good provides the foundation for liberty to take its place as the exclusive locus of value in conduct, setting the stage for modernity to erect institutions of freedom, liberating the future from the past. Yet can liberty escape the pitfalls of a highest good if it confers legitimacy upon any end that is willed in accord with its privileged form of volition? Whether interpreted empirically as the freedom of desire that utilitarianism advances or conceived a priori as the formal willing of social contract construction, liberty has a given content of its own, defined in function of the self as a natural endowment that precedes all conventions. Can its given agency command exclusive legitimacy any more than any given end?

If ends can no longer be rationally ordered, then every desire and disposition stands on a par, suggesting that nothing can favor one action, character, or institution over another except a greater quantity of resulting desire satisfaction. Yet if desires are qualitatively different, contingent and variable, and therefore devoid of any necessary harmony, a utilitarian calculus can hardly be applied, let alone dispel the nihilist challenge of why desire satisfaction should override any competing consideration. Lacking the universality that would permit the desire satisfaction of one individual to be realized compatibly with that of others, the liberty of desire simply cannot fit the form of right and occupy the position of a privileged ethical principle.

To enjoy the universal entitlement of an archimedean right, liberty must abstract from the particularity of its empirical ends and obtain a lawful form guaranteeing the harmony of its realizations. What better meets this requirement than the agency of property relations, where individuals invest their will in an external factor that can be recognized as the receptacle of their volition because it does not intrude upon the respective domains in which others have laid their own wills? In determining oneself as an owner, what counts is not the desire satisfaction that ownership brings oneself or others, but simply the achievement of a recognized embodiment of will that stands in harmony with the analogous embodiments of other owners. Unlike pursuing one's desires, exercising one's freedom of ownership, be it by staking an original ownership claim, using one's property or transferring it by contract, involves willing in a manner that not only automatically coexists with the ownership rights of others, but operates prior to the institution of any further forms of community. For this reason, the right of contract can figure as a foundation for ethical construction, occupying a primordial state of nature preceding all legitimate convention from out of which institutions can gain sanction as contractually chosen instruments for upholding property right. This familiar formula for social contract theory inaugurates the liberal challenge to pre-modern tradition, setting universal property right as a revolutionary principle overturning all conventions that restrict the freedom of contract.

Yet why should the formal right of property owners reign supreme? Taken as a principle of ethical construction, property right falls into the very trap of foundationalism that waylaid the quest for a highest good and put liberty at center stage. To play the archimedean role of foundation of justice, property right must lay claim to an antecedently given content that somehow warrants the privilege to confer validity upon whatever conventions can be derived from it. Although property right does not consist of a given end, it still presents a form of willing with just as much a given content as any highest good. Consequently, the form of its agency does not derive from the determining procedure for which it is put to constructive use. Yet if legitimacy consists in being the outcome of its choice procedure, the contracting will cannot itself enjoy the validity it alone confers. Not only does it determine derivative institutions different from itself, but these grounded conventions owe their design not to themselves but to the prior choice procedure whose own authority remains an immediate stipulation. In both respects, the Enlightenment project of procedural ethical construction remains caught in the foundational categories of positing, of positor and posited, of ground and grounded, failing to account for the self-determination with which modernity distinguishes itself from pre-modern tradition and postmodern irony. Neither the collapse of an ethics of a highest good nor the incompatibilities of the freedom of desire can sanction the formal volition of social contract and its procedural ethics. Their failings certify only their own inadequacy, leaving intact the challenge of nihilism and the abiding problem of justifying a principle of willing that founds without being founded.

The modern quest for beauty appears to suffer a similar fate, if the Enlightenment be allowed to provide the measure for modernity. Whereas pre-modern thought sought beauty in mimetic congruence with reality, Enlightenment aesthetics rejects founding beauty in natural form since mimesis either reduces beauty to a redundant copy, erasing the distinction between the prosaic and the beautiful, or condemns beauty to a deficient replica, whose deviating individuality is deprived of any aesthetic worth. Instead, Enlightenment thought turns to the process of reception to indirectly determine aesthetic worth, just as its foundational epistemology constructed objectivity from structures of knowing and its procedural ethics conjured right from social contract.

If aesthetic experience be construed in empirical terms, aesthetics can only provide standards of beauty by appealing to some common psychological nature that allows all viewers to respond with comparable pleasure to the same appearances. Yet, like empirical desire, an empirical taste for beauty cannot command the coherent universality norms demand. Appeal to human psychology either reduces aesthetic response to an instinctual necessity from which normativity disappears or to a provisionally observed pattern of response to which necessity can no more be ascribed than can juridical privilege. If alternately the empirical process of reception be defined by the cultural conventions of language or of some institutionalized "artworld" analogous problems persist. If aesthetic worth be measured by what those conventions posit as art or beauty, the question remains as to why the favored objects qualify, a question that has no answer if objective features are not the locus of aesthetic worth and the defining conventions operate without any other identifiable grounds.

Moving away from empirically given psychological or conventional factors to formal, a priori features of aesthetic experience might seem to offer the same benefits as the move from psychologism to transcendental constitution in epistemology and the move from utility to property right in liberal ethics. Aesthetic standards might now command the universality and necessity required, even if both be rooted in the subjective or intersubjective character of reception, rather than in the objective features of the object of beauty. Yet so long as beauty be founded in the determining process of reception, all objects owe their aesthetic value to the same foundation, reducing beauty to a universal property that is fundamentally indifferent to the individuality that distinguishes the originality of fine art and the creativity of artistry from the common form of artifacts and the rote technique of craft. Because Enlightenment aesthetics determines beauty by a common process of reception given independently of the individual appearances to which beauty is ascribed, all chance is lost of capturing the unity of meaning and configuration, the inseparability of sensuous detail and significance that is the counterpart of the creativity and originality in fine art. Just as autonomy is forsaken if validity derives from a choice procedure preceding the conduct it sanctions, so the individuality of beauty falls beyond grasp if the standard of taste lies in a separate ground, be it a form of reality to be imitated by every work with the same topic or a structure of reception common to every appearance judged to have aesthetic value.

At each turn, in knowledge, in conduct and in beauty, the Enlightenment project fails to escape the dilemmas of tradition by failing to overcome foundations in theory, practice and aesthetic judgment. By substituting construction for contemplation, liberty for a highest good and a critique of taste for mimetic fidelity, Enlightenment thought still remains within the foundational orbit that undermines all its efforts to provide a viable justification for modernity.

For post-modernism, this result signifies much more than that the Enlightenment's attempt to provide foundations for autonomous reason is a hopeless endeavor. Post-modernism further concludes that modernity's project of advancing autonomy in reason, conduct and art is equally quixotic. To come to this conclusion by exposing the dilemmas of epistemological foundationalism, procedural justice and the critique of taste only makes sense if the autonomy of reason, conduct and beauty cannot be justified without appeal to foundations. The failure of the Enlightenment can only serve as an indictment of modernity if post-modernism further presumes that reason, conduct and art cannot escape the hold of foundations. Rejecting the possibility of radical independence, of liberating the future from the past, post-modernism accordingly maintains that thinking, action and artistry can only take their normative bearings from privileged vocabularies that foundations provide. Because, however, post-modernism understands that the appeal to foundations can never achieve self-legitimation, post-modernism recognizes that these foundations must lack the features that justification would require. Instead of being universal, necessary and conceptually transparent, the privileged foundations are actually particular, contingent and opaque to reason. Consequently, whatever norms are grounded upon them are merely values, that is, relative, perspectival ideals whose claim to universality can only really represent an attempt by their particular authors to impose their will upon all others.

Thus, even though post-modernism joins pre-modernity in accepting the hegemony of foundations over human affairs, post-modernism rejects the possibility of endowing any one tradition with privileged authority. Recognizing instead that all foundations are particular and arbitrary, post-modernism can offer only one recipe for a civilization of its own: in knowing, in action and in art, a consistently post-modern community will advance norms that are openly particular and arbitrary. Instead of conforming to pre-modern tradition or following the universal aspirations of modernity's embrace of freedom, post-modern civilization will mold itself upon a forthright will to power, oppressing without veils or excuses. The values that post-modernity advances will give expression to the past as embodied in the given particular character of its agents. Yet because epistemological claims are made with the same perspectival arbitrariness as ethical and aesthetic values, post-modernity must treat the past as an instrument for advancing the triumph of the consistent will to power that it opposes to the alleged illusion of self-determination and its universal modern rights.

Post-modernity may well regard all foundations and their derivative values with an irony befitting their historical contingency and relativity. Yet ironic detachment is not an invitation to multicultural toleration or democracy, as irresolute post-modernists would like to believe. Acknowledgment of the arbitrariness of all values provides no rationale for toleration, for toleration is itself a particular ideal to which irony should be indifferent. If anything, an active opposition prevails, for toleration and democracy rest upon a very firm commitment to the equal right of each culture and political agent, a commitment that can only be realized through the internal reformation or external curbing of incompatible, intolerant value systems. To find consistent expression, post-modern irony would better pursue a frankly perspectival world order, where the arbitrary will of a single unconstrained leader imposes its domination for the few with whom it shares some given particular identity.

The Dilemma behind the Rhetoric of Post-modernism

So long as modernity finds no better defenders than the foundational thinkers of the Enlightenment, post-modernism can readily achieve a rhetorical triumph, identifying the legitimation of modernity with the Enlightenment project, unmasking the foundational dilemmas of the latter, and deconstructing the will to power allegedly underlying and subverting the universalist claims of modern civilization. Yet even in this endeavor, post-modernism cannot free itself of the very duplicities it condemns. To supersede pre-modernity and modernity alike, post-modernity must deny the autonomy of reason, conduct and art, and uphold the hegemony of historically contingent foundations over all the value-positing activities distinguishing humanity. If, however, post-modernism's diagnosis of the conditioned character of theory and practice is true, how can it conform to its own verdict? If all knowing and doing is conditioned by unjustifiable conventions, how can post-modernism be more than a perspectival ideology whose claims are themselves relative to its own contingent foundations? In order to legitimate its global judgment upon discourse and conduct against competing views, post-modernism must transcend the hegemony of contingent foundations, contradicting its own thesis. If post-modernism instead admits to being one ideology among others, it can hardly present its characterization of modernity as more than a strategic tool of its own self-assertion. Even if the Enlightenment project collapses when subjected to self-critique, post-modernism cannot justifiably maintain that modernity is to be measured by the Enlightenment and that autonomous reason cannot be given an internally consistent account and a coherent realization.

Modernity beyond Enlightenment and Post-modernism

What post-modernism never even considers is the key to modernity's break with tradition, a key that the Enlightenment equally ignores. This is the revolutionary insight that self-determination does not and cannot have foundations. Unless freedom's independence from foundations can be comprehended, the normativity of truth, right and beauty cannot be redeemed from the dogmas that have cast them under a cloud of suspicion. Neither pre-modern tradition, nor Enlightenment thought, nor post-modernism can absorb this crucial comprehension so long as they retain their common belief that the logic of justification is identical to the logic of foundationalism. Each strikes a different attitude to the identity they all acknowledge, pre-modernists trusting the authority of a particular foundation, Enlightenment thinkers advancing foundations that are privileged determiners rather than privileged givens, and post-modernists acquiescing to the ubiquity of foundations while acknowledging the historical contingency of each one. Amidst this diversity, none dares wager that justification could avoid appeal to some foundation or other.

Far easier is recognizing that justification is always hobbled so long as it rests upon some antecedent ground, conferring validity upon something other than itself. What could be more absurd than looking for the basis of truth in something outside the truth, in searching for right in what is not right itself, in locating beauty in what itself lacks all aesthetic worth? Certainly any recourse to foundations warrants sceptical detachment, but is post-modern irony any less naive in its acceptance of the hegemony of foundational justification? An anti-foundationalism that decries truth, right and beauty by unmasking the contingency of all privileged paradigms is an anti-foundationalism in name only. Not only does such an impostor offer its own grand narrative of the global conditioning of reason, conduct and artistry by historical conventions, but it lacks the candor to admit the incoherence of its own unconditioned claims about theory and practice.

The authentic alternative to foundations is so close at hand in the dilemmas of foundationalism, so radical in its break with tradition, so potent in paralyzing the ideological challenge of fascism, and so pervasive in a modern age that the simplicity of its solution risks making it virtually invisible. A civilization emerging from oppressive tradition, weaned on the Enlightenment misconception of modernity, and tempted by post-modern irony is thoroughly accustomed to see no justification if foundations are not advanced. What appears without ground is assumed to be arbitrary, a posit of a will that conceals its founding authorship. A truly groundless content is simply beyond comprehension for a viewpoint blind to validity without foundations. Yet, seeing what foundational thinking finds invisible is precisely the challenge on which the justification of modernity hangs.

Self-determination is inconceivable in foundational terms. To qualify as self-determined, something must be what it has determined itself to be. The character of self-determination cannot be something immediately given, for any given content would not have resulted from self-determination, but be an endowment deriving from some other source. Whatever content self-determination possesses must instead arise in the process by which the self-determined self engenders itself as free. Accordingly, freedom cannot have any grounds, but must rather ground itself in the process constitutive of self-determination.

The relation between normativity and self-determination resides in this independence from external grounding that has its positive expression in the self-grounding character of freedom. The internal logic of foundational argument already prepares the way for this connection without need of appeal to any competing truths. If a foundation confers validity, the foundation cannot possess the validity it provides unless it rests upon itself just as does what it founds. The foundation lacks validity unless it becomes self-grounding, transforming itself from a foundation both different and antecedent to what it founds into a subject that determines itself. Applied to the three domains of normativity, the outcome of foundationalism signals that truth must rest upon itself, that right must be defined by nothing but right, and that beauty must be autonomous.

These three injunctions lose their tautological appearance the moment autonomy in reason, conduct and art get developed in theory or confirmed as defining keystones of modern practice.

Modernity and Reason

Modernity can no more do without autonomous reason than reason can do without autonomy. Pre-modern philosophy was content to seek the principle of reality without worrying about what came first in thinking. Yet, if each candidate for first principle of ontology had to have a non-derivative content, nothing could certify the supremacy of one alternative over another. Enlightenment thought turned its attention to what came first in thinking, as if locating some privileged ground in the process of knowing could overcome ontological relativity. Either strategy puts philosophy in the same bind: the moment philosophy begins with a foundation, the character and primacy of the privileged term cannot help but be taken for granted as something decided before and, therefore, outside of philosophical investigation.

Philosophical reason thereby condemns itself to heteronomy. Instead of accounting for all its own terms, philosophy allows itself to be ruled by contents it has not justified, undermining any hope of reason legitimating itself and serving as the ultimate tribunal for modernity's break with the past. If philosophy is not to take its own method and subject matter for granted, but treat its form and content as philosophical problems to be decided within philosophy, reason must operate without any preconceptions concerning what it should be addressing or how it should proceed. Yet in that case, philosophy must begin without any initial claims about reality or knowing, that is, without any determinate foundations. To do so, philosophy proper must begin without any determinate method or content, and if philosophy is to move beyond being an empty word to arrive at any determinate claims in any determinate order, both the content and its form of presentation must develop without reliance upon any external factors. Philosophical thought must therefore be self-determining, giving itself a content through its own efforts whose form must equally be the product of philosophical reason. To escape dogmatism, philosophy must overcome foundations by developing as an autonomous reason. Moreover, because philosophy cannot begin with the assumption of any determinate ground, the self-determination that philosophical reason must realize can be nothing but self-determination per se, rather than the autonomous development of some given substrate. The form of philosophical truth turns out to be identical with the logic of self-determination.

This revolutionary outcome might appear hopelessly at odds with common understandings of logic, given how traditional and Enlightenment thought have treated logic as a formal deductive discipline, whose rules of inference are axiomatic givens applying not to themselves but to extraneously given subject matters. Yet, the very project of logic cannot possibly meet success unless it converges with the same identification of validity and autonomy that emerges from philosophy's grappling with foundations. As a thinking of thinking, logic begins with a unity of method and content, where knowing and its object are indistinguishable. If logic is to prescribe what is valid thinking, the thinking that logic employs must be identical with valid thinking itself. Logic would, however, be begging its question if it started out with either a determinate method or a determinate content. If logic is to establish the form and content of valid thinking, these must both be the result, rather than the presupposed starting point of its labors. Whatever is established within logic must then develop itself, for nothing outside of logical investigation can legitimately furnish any material or procedure. Since logic must begin without any givens, the autonomous subject matter it ends up presenting can be nothing other than self-determination, once more confirming how the legitimacy of reason depends upon supplanting reliance upon foundations with autonomy.

Precisely *how* logic thinks valid thinking and *how* philosophy presents the autonomous development of self-determination are matters that lie beyond the identification of normativity and freedom. Although these questions have received little attention outside of Hegel's much maligned *Science of Logic*, whether any modern thinker has yet satisfactorily answered them has no bearing upon the possibility of doing so or upon the possibility of legitimating modernity. What can be said without reserve is that if philosophy remains caught within the foundationalisms of traditional ontology or Enlightenment epistemology, reason can never achieve the radical self-responsibility and self-legitimation of a genuinely autonomous thinking. But if philosophy overcomes foundations by thinking autonomously, modernity can possess the self-legitimating reason it needs to break with tradition, disarm the post-modern challenge and liberate the future of thought from the past while rescuing the past from the ideological designs for tomorrow.

Modernity and Right

What holds for theory, holds all the more palpably for practice. Just as truth must ground itself, so what is right must break with foundations and give itself its own order. Unless conduct is self-determined, unless institutions comprise the reality of freedom, the normativity of practice will lie in external factors whose own legitimacy remains always in question. Yet how can conduct escape the hold of historical conditions and achieve self-determination? Is modernity's whole project to overthrow traditional oppression with universal institutions of freedom an illusory dream?

Enlightenment ethics already raised the banner of liberty as the only viable alternative to the traditional pursuit of a highest good through virtue and character. Yet because Enlightenment ethics treated liberty as a foundation for ethical construction, be it by employing a utilitarian calculus or social contract, self-determination got conflated with the structure of positing, where what determines is different from what gets determined. This reduction precludes any possibility of coherently conceiving freedom, for so long as the will is thought to have a given nature preceding its volition, it can never exhibit the reflexivity of autonomy, where what determines and what gets determined are one and the same. Self-determination easily appears to be an utter impossibility – first, because no willing can take place without a "natural" agency endowed with a capacity of choice preceding each and every act of will, and secondly, because no will can determine its own character by itself without falling into the paradox of having to act upon itself with an agency that is supposed to issue from its own activity. Because of just this difficulty of conceiving agent and patient as one and the same, Plato early on abandoned all thought of self-determination, concluding instead that the soul must be divided into a ruling, rather than self-ruling, component, lording over a subject accessory, just as the polis would have to be divided into a guardian class ruling over a subject class.

What resolves these difficulties is emancipating the conception of freedom from the parameter of the single subject, a parameter that locked Enlightenment thought into conflating self-determination with positing, the determining of something by an antecedently given determiner. If freedom is conceived not as a function of the self but as a conventional structure of right in which a plurality of agents interact in function of universal entitlements, the reflexivity of self-determination becomes a possibility not so unfamiliar for a modern age.

To be self-determined, an agent must determine not only what ends are willed, but who does the willing. Instead of being merely a natural endowment preceding volition, the agency of self-determined conduct must have a character that is determined by its own willing. Autonomous agency must thus be artificial or conventional, rather than natural. Although an isolated agent can never act without presupposing, rather than determining, its own given agency, the will becomes able to determine its own agency by participating in an enacted context of rights.

Self-government, the most sovereign such context, shows perhaps most clearly how self-determination involves a conventional agency that only the interaction of rights can provide. To exercise political self-determination, no natural endowments or functions of the self suffice. Although one cannot engage in self-government unless one possesses the natural, psychological and linguistic capabilities making possible independent action and respect for the rights of others, political self-determination is impossible unless one belongs as a recognized citizen to an existing state containing the institutions of self-government. For any individual to enjoy this membership, the other citizens must simultaneously exercise their political self-determination simply because the reality of self-government consists in nothing but the political activities of self-governing citizens. Moreover, each citizen must enjoy equal political opportunity, for unless the right to political self-determination is universal, the politically privileged will lord over the rest, transforming self-government into a regime where a ruling elite imposes order upon a subject class. Because the enabling institutional context of political freedom is itself determined by the exercise of self-government, political self-determination is able to do what the individual will can never independently achieve: namely, determine not only the content of its volition but the nature of its agency. By performing their political role, citizens determine their own artificial, conventional agency as self-governing individuals and they do so by exercising a political right whose guaranteed exercise is part and parcel of the reality of self-government. Consequently, political freedom has normativity built into it. Self-government, like every other mode of self-determination, is itself a structure of right in which individuals impose upon themselves an entitled agency, whose conventional autonomy cannot be exercised apart from mutual recognition of its validity. Whereas Enlightenment ethics treated freedom as a principle from which valid institutions are derived, the interaction of self-determination renders freedom the reality rather than a foundation of justice.

The resulting identity between self-determination and normativity in conduct might seem vain if, first, a plurality of rights calls for restraining self-determination in any of its institutional modes and, second, if the existence of the institutions of freedom is conditioned by natural, psychological and historical factors.

The first question bears directly upon the legitimacy of political freedom, for if there be non-political rights, self-government must be prevented from overriding those entitlements, a task that seems to call for a restraining power undermining political sovereignty, as well as the exclusive validity of self-determination. The problem disappears, however, as soon as one recognizes that political freedom cannot be exercised unless non-political freedoms are simultaneously realized. In the first place, individuals can hardly claim political freedom, let alone any other right, unless they all have determined themselves as owners, enjoying recognition as proprietors of at least their own bodies, without which they remain subject to enslavement. Similarly, unless agents possess the right of moral accountability, holding one another responsible only for what they have done on purpose and only for those ramifications of their actions that are prefigured in their motives, they cannot enjoy the independent political responsibility on which self-government depends. By the same token, unless the family is transformed into an institution of freedom, where spouses codetermine household affairs independently of any other kin domination, subordinate family members can hardly enjoy equal access to society and state and participate in politics on a par with others. And without the realization of social freedom through a civil society enabling individuals to choose their occupation and needs in the market under equal protection of civil law and public welfare guarantees of equal economic opportunity, the barriers of social privilege and the fixity of tradition will block political opportunity for the socially oppressed. Self-government therefore need not be subject to the anomaly of a higher authority to uphold pre-political right. A higher foundation is precluded, for unless the rights of property, moral, family and social freedom are all enforced, political self-determination cannot itself exist. Hence, far from comprising a restriction upon political freedom, the non-political freedoms are necessary conditions of its own realization. Not only does self-rule require the establishment of property rights and moral, household and social autonomy, but political freedom cannot sustain itself unless all these non-political freedoms are upheld by the state in conformity with the practice of self-government. The reality of freedom therefore comprises a self-sustaining system of rights, grounding itself through the sovereign autonomy of self-governing citizens, whose own political self-determination must uphold the totality of their freedom in order to subsist.

Admittedly, the institutions of freedom cannot maintain themselves unless a whole variety of natural, psychological, and historical conditions are at hand. The physical, chemical and biological requirements of rational agency must be present, just as must the cultural developments that permit individuals to interact in recognition of their rights. Yet these *enabling* conditions cannot play the role of *juridical* conditions, as foundationalists would like to believe. Precisely because the physical, psychological and linguistic conditions of rational agency underlie each and every form of convention, they can hardly provide a basis for distinguishing between the just and unjust institutions, let alone the true and false theories, that they equally make possible. For this reason, the contingency of the enabling conditions of free conduct in no way limits the normative independence of self-determination. The practical project of modernity may be frustrated in a particular time and place by natural conditions beyond rational control, such as climatic calamities, asteriod collisions, or solar burnout. Further, contingent cultural developments can always prevent freedom from prevailing over oppression. Nevertheless, wherever and whenever physical nature is permitting, the mere presence of a tradition signals the possibility of struggling for emancipation.

This struggle is not without its complications, given how the genesis of a modern civilization is determined by the very structure of the institutions of freedom. Although those institutions are defined independently of all foundations, and therefore independently of all local tradition, they need not and, indeed, cannot come into being with a reach fitting their inherent universality. The founding of the institutions of self-government is not an exercise of political freedom, nor a merely political deed, for it equally involves the whole series of historical developments by which universal property right becomes recognized, moral autonomy triumphs over sacred and secular tutelage, household hierarchy is overthrown, and social bondage is overcome. Since these transformations are subject to all the contingencies of history, the first modern community cannot help but emerge in a particular time and place, bearing the contingent stamp of its regional baptism. Even though modernity may command universal validity because its institutions of freedom are independent of each and every heritage, the local character of the genesis of modernity creates a situation where a first modern civilization arises in face of pre-modern communities that have not as yet modernized themselves. The transformation of these abiding traditional communities into modern nations may then take the form of a regional assimilation, which may or may not involve direct political domination. To take our planetary experience as an example, if modernity has truly first arisen in the West, then modernization will appear as Westernization, especially if the first modern civilization still bears the mark of contingent cultural features that are incidental to the institutions of freedom. Whatever the resulting scenario, the normativity of freedom sets the standard for what should be done. Pre-modern tradition as well as postmodern tyranny ought to give way to the institutionalization of freedom, which by its very nature cannot come to fruition without the voluntary involvement of those who are to be emancipated. Consequently, the modernization following upon the emergence of the first modern nation has no other legitimate destination than the universal political independence that eliminates all trace of colonial and neo-colonial domination.

Modernity and Beauty

Although modern recognition of the normativity of freedom does not preclude controversy from raging over how autonomous thought actually determines itself and what specific content defines the different institutions of freedom, the situation of fine art in modernity raises questions of a different sort. Aesthetics has always confronted the problem of how beauty can persist untarnished through the ages whereas past achievements in theory and practice lose their lustre when judged by the modern standard of autonomous reason. In some respects the artistic achievements of former times seem to present no longer attainable heights and the whole place of beauty in modern culture seems to betray a diminished stature compared to past ages where right and truth had no precedence over the beautiful.

Resolving the post-modern challenge to aesthetic worth leaves an outcome of ambiguous meaning for art in modernity. The inability of either mimetic theory or the Enlightenment critique of taste to capture the individuality of beauty and comprehend the originality and creativity of fine art can be overcome by casting aside external foundations and locating the focal point of beauty in the unity of meaning and sensuous form. This unity allows the individual detail of an aesthetic appearance to retain essential significance, while placing the normativity of beauty in the self-sufficient and original achievement by which a work presents a meaning worthy of universal attention in a sensuous configuration uniquely suited to that significance. Unless an object of beauty can convey something of universal significance, it forfeits the breadth of appeal that allows aesthetic worth to be appreciated in any age and culture. And unless the object of beauty conveys something universally significant in a uniquely adequate shape, it sacrifices its standing as something that deserves individual attention. The normativity of freedom provides the very possibility of universally significant meanings by redeeming the search for truth, be it sacred or profane, the struggle for right, and art's own quest for beauty as affairs of fundamental meaning to which no rational agent can be irrevocably indifferent. Consequently, the meaning fit for beauty will enable art to provide an essential self-understanding for humanity, so long as the meanings in question are susceptible to sensuous configuration. Because philosophical argument is purely conceptual, whereas secular and sacred conduct has a sensuous dimension, it can be no accident that fine art has preeminently grappled with configuring the essential problems of religion and ethics. Since, however, pre-modern, modern and post-modern civilizations view the normativity of these configurable domains in different ways, their aesthetic options differ.

Beauty might seem most amenable to pre-modern realization since pre-modern civilization finds legitimacy in the contingent existence of given tradition, which can only be captured by non-conceptual representation. Insofar as traditional values cannot be conceptually comprehended, artistic presentation would provide a better vehicle for pre-modern self-understanding than philosophical theory. Moreover, since the fundamental meanings to be expressed in traditional art cannot be properly communicated by thought alone, the images of art would seem especially suited to fit what pre-modern civilization finds fundamentally important. Not only would beauty here have special cultural significance as a privileged revelation of ultimate values, but the aesthetic unity of meaning and configuration would be achievable to supreme degree.

By contrast, modernity presents fine art with a more formidable challenge. Although subscribers to the modern spirit can look backwards in appreciation of how traditional art delivers supremely apt expressions of pre-modern self-understanding, traditional aesthetic forms can no longer give adequate artistic configuration to the revolutionary values of rational autonomy. Once validity gets identified with self-determination no given sensuous configuration can sufficiently convey what is of ultimate significance. Whereas traditional norms can be better captured by exemplary individuals than by abstract principle, the unconditioned universality of freedom is more suitably expressed by conceptual thought than by the image of any example.

Art that passes itself off as post-modern by making ironic use of traditional aesthetic forms is really following but one of the modernist strategies that artists may employ in giving aesthetic expression to the modern self-understanding. Whereas post-modernism forsakes freedom for the will to power in sober homage to the hegemony of foundations, "post-modern" art playfully recycles pre-modern styles and genres, quoting past examples of beauty with subjective license as if to show that no given configuration is adequate to the free creativity that rules supreme. Post-modern architecture, perhaps the most renowned exponent of this gambit, might appear polemically at complete odds with the pure abstractions of modernist "international style" construction. Nonetheless, the post-modern architectural quotation of tradition manifests the same withdrawal of authority from past canons as the modernist invention of expressly non-natural and non-traditional purities of design. The purely artificial system of serial music that brings constructed order to the arbitrary solitude of expressionist atonality is connected by the same unity of spirit with the expressly non-traditional recycling of primitive rhythms and folk melodies by early Stravinsky, Bartok, and the minimalists, as well as with the neo-classical recasting of classical forms in departure from traditional harmonic development. In each case, composers seek a musical expression for a creative autonomy that no given form can preeminently satisfy. Similarly, whether modern painting opts for pre-modern delineation and coloration with expressionist abandon, the mechanical naturalism of photo-realism, the free associations of dada and surrealism, completely non-objective abstraction, or conceptual art's self-annulling retreat from image to self-reflection, the same aesthetic modernism is at work seeking to exhibit how no sensuous given can have any privileged rule over the autonomy that alone should count. The liberation from the past could not be sharper. Epic literature could provide a ready vehicle for a closed world of tradition, where individual striving has significance only within the completed frame of a national saga, where everything that appears is essential and nothing merely inward is more than personal eccentricity or madness. The modern self-understanding needs a novel form, one in which the self-determination of the individual sets the course of development, where the yearnings of inner life demand satisfaction over and against the contingencies of an indifferent outer world, where classical idealizations get replaced by encounters with any situation high or low and any character regal or depraved that an open future can bring.

Modernity may not condemn art to death, but it has raised an aesthetic dilemma with no easy solutions. Because modern art must show how no given configurations can accord sufficient shape to autonomy, modern art risks displaying an ultimate indifference to the form and content of its images. Since aesthetic worth resides in the congruence between the configuration and fundamental meaning in the work of art, such indifference threatens to undermine the stature of modern art's own creations. For this reason, irony can be an emblem of modern art even if the conceptual adequacy of autonomous reason and autonomous conduct shields truth and right from post-modern irony.

Despite this disparity in the modern predicament of beauty on the one hand, and of truth and right on the other, the consequences for past and future remain the same. Just as the autonomies of reason and conduct make modern thought and practice a constitutively global imperative, transcending the limits of any regional heritage, so the challenge facing modern art is the summons of a universal culture, whose aesthetic productions must configure a freedom of spirit that no tradition can confine. A nostalgic multiculturalism may reanimate a museum of folklore and rote tradition, while a post-modern tyranny may banish or mummify the exceptions to its own artistic purities. A modern art, by contrast, is a genuine world art, no matter what particular languages and iconographies it employs.

Whether modernity marches on to a consistent global realization, whether traditional reaction or post-modern counterrevolution rejuvenates the rule of foundations, whether natural or human catastrophes banish intelligent life from the planet, and whether these same options play themselves out in other galaxies are contingent matters whose indeterminacy is guaranteed by the freedom of the future from the past. Yet the contingency of the triumph of modernity is no blemish on the legitimacy of the modern project. Precisely because self-determination depends on no foundations, the normativity of freedom always retains authority, forever extending validity to modernity, forever emancipating the future of truth, right and beauty from the hold of the past, and forever enabling the truth of the past to be free of interests in the future.

Index